Morse Theoretic Aspects of p-Laplacian Type Operators

Mathematical Surveys and Monographs

Volume 161

Morse Theoretic Aspects of p-Laplacian Type Operators

Kanishka Perera

Ravi P. Agarwal

Donal O'Regan

American Mathematical Society

Providence, Rhode Island

2000 *Mathematics Subject Classification.* Primary 58E05, 47J05, 47J10, 35J60.

For additional information and updates on this book, visit
www.ams.org/bookpages/surv-161

Library of Congress Cataloging-in-Publication Data

Perera, Kanishka, 1969–
Morse theoretic aspects of p-Laplacian type operators / Kanishka Perera, Ravi Agarwal, Donal O'Regan.
p. cm. — (Mathematical surveys and monographs ; v. 161)
Includes bibliographical references
ISBN 978-0-8218-4968-2 (alk. paper)
1. Morse theory. 2. Laplacian operator. 3. Critical point theory (Mathematical analysis) I. Agarwal, Ravi, 1947– II. O'Regan, Donal. III. Title

QA614.7.P47 2010
515′.3533–dc22 2009050663

∞ The paper used in this book is acid-free and falls within the guidelines
established to ensure permanence and durability.
Visit the AMS home page at http://www.ams.org/

10 9 8 7 6 5 4 3 2 1 15 14 13 12 11 10

Contents

Preface

The p-Laplacian operator

$$\Delta_p u = \operatorname{div}\left(|\nabla u|^{p-2}\,\nabla u\right), \quad p \in (1, \infty)$$

arises in non-Newtonian fluid flows, turbulent filtration in porous media, plasticity theory, rheology, glaciology, and in many other application areas; see, e.g., Esteban and Vázquez [**48**] and Padial, Takáč, and Tello [**90**]. Problems involving the p-Laplacian have been studied extensively in the literature during the last fifty years. However, only a few papers have used Morse theoretic methods to study such problems; see, e.g., Vannella [**130**], Cingolani and Vannella [**29, 31**], Dancer and Perera [**40**], Liu and Su [**74**], Jiu and Su [**58**], Perera [**98, 99, 100**], Bartsch and Liu [**15**], Jiang [**57**], Liu and Li [**75**], Ayoujil and El Amrouss [**10, 11, 12**], Cingolani and Degiovanni [**30**], Guo and Liu [**55**], Liu and Liu [**73**], Degiovanni and Lancelotti [**43, 44**], Liu and Geng [**70**], Tanaka [**129**], and Fang and Liu [**50**]. The purpose of this monograph is to fill this gap in the literature by presenting a Morse theoretic study of a very general class of homogeneous operators that includes the p-Laplacian as a special case.

Infinite dimensional Morse theory has been used extensively in the literature to study semilinear problems (see, e.g., Chang [**28**] or Mawhin and Willem [**81**]). In this theory the behavior of a C^1-functional defined on a Banach space near one of its isolated critical points is described by its critical groups, and there are standard tools for computing these groups for the variational functional associated with a semilinear problem. They include the Morse and splitting lemmas, the shifting theorem, and various linking and local linking theorems based on eigenspaces that give critical points with nontrivial critical groups. Unfortunately, none of them apply to quasilinear problems where the Euler functional is no longer defined on a Hilbert space or is C^2 and there are no eigenspaces to work with. We will systematically develop alternative tools, such as nonlinear linking and local linking theories, in order to effectively apply Morse theory to such problems.

A complete description of the spectrum of a quasilinear operator such as the p-Laplacian is in general not available. Unbounded sequences of eigenvalues can be constructed using various minimax schemes, but it is generally not known whether they give a full list, and it is often unclear whether different schemes give the same eigenvalues. The standard eigenvalue sequence based on the Krasnoselskii genus is not useful for obtaining nontrivial critical groups or for constructing linking sets or local linkings. We will work with a new sequence of eigenvalues introduced by the first author in [**98**] that uses the $\mathbb{Z}_2$-cohomological index of Fadell and Rabinowitz. The necessary background material on algebraic topology and the cohomological index will be given in order to make the text as self-contained as possible.

One of the main points that we would like to make here is that, contrary to the prevailing sentiment in the literature, the lack of a complete list of eigenvalues is not a serious obstacle to effectively applying critical point theory. Indeed, our sequence of eigenvalues is sufficient to adapt many of the standard variational methods for solving semilinear problems to the quasilinear case. In particular, we will obtain nontrivial critical groups and use the stability and piercing properties of the cohomological index to construct new linking sets that are readily applicable to quasilinear problems. Of course, such constructions cannot be based on linear subspaces since we no longer have eigenspaces. We will instead use nonlinear splittings based on certain sub- and superlevel sets whose cohomological indices can be precisely calculated. We will also introduce a new notion of local linking based on these splittings.

We will describe the general setting and give some examples in Chapter 1, but first we give an overview of the theory developed here and a preliminary survey chapter on Morse theoretic methods used in variational problems in order to set up the history and context.

An Overview

Let Φ be a C^1-functional defined on a real Banach space W and satisfying the (PS) condition. In Morse theory the local behavior of Φ near an isolated critical point u is described by the sequence of critical groups

$$(1) \qquad C^q(\Phi, u) = H^q(\Phi^c \cap U, \Phi^c \cap U \setminus \{u\}), \quad q \geqslant 0$$

where $c = \Phi(u)$ is the corresponding critical value, Φ^c is the sublevel set $\{u \in W : \Phi(u) \leqslant c\}$, U is a neighborhood of u containing no other critical points, and H denotes cohomology. They are independent of U by the excision property. When the critical values are bounded from below, the global behavior of Φ can be described by the critical groups at infinity

$$C^q(\Phi, \infty) = H^q(W, \Phi^a), \quad q \geqslant 0$$

where a is less than all critical values. They are independent of a by the second deformation lemma and the homotopy invariance of cohomology groups.

When Φ has only a finite number of critical points $u_1, \ldots, u_k$, their critical groups are related to those at infinity by

$$\sum_{i=1}^{k} \operatorname{rank} C^q(\Phi, u_i) \geqslant \operatorname{rank} C^q(\Phi, \infty) \quad \forall q$$

(see Proposition 3.16). Thus, if $C^q(\Phi, \infty) \neq 0$, then Φ has a critical point u with $C^q(\Phi, u) \neq 0$. If zero is the only critical point of Φ and $\Phi(0) = 0$, then taking $U = W$ in (1), and noting that Φ^0 is a deformation retract of W and $\Phi^0 \setminus \{0\}$ deformation retracts to Φ^a by the second deformation lemma, gives

$$\begin{aligned} C^q(\Phi, 0) = H^q(\Phi^0, \Phi^0 \setminus \{0\}) &\approx H^q(W, \Phi^0 \setminus \{0\}) \\ &\approx H^q(W, \Phi^a) = C^q(\Phi, \infty) \quad \forall q. \end{aligned}$$

Thus, if $C^q(\Phi, 0) \not\approx C^q(\Phi, \infty)$ for some q, then Φ has a critical point $u \neq 0$. Such ideas have been used extensively in the literature to obtain multiple nontrivial solutions of semilinear elliptic boundary value problems (see, e.g., Mawhin and Willem [**81**], Chang [**28**], Bartsch and Li [**14**], and their references).

Now consider the eigenvalue problem

$$\begin{cases} -\Delta_p u = \lambda |u|^{p-2} u & \text{in } \Omega \\ u = 0 & \text{on } \partial\Omega \end{cases}$$

where Ω is a bounded domain in $\mathbb{R}^n$, $n \geqslant 1$,

$$\Delta_p u = \operatorname{div}\left(|\nabla u|^{p-2} \nabla u\right)$$

is the p-Laplacian of u, and $p \in (1, \infty)$. The eigenfunctions coincide with the critical points of the C^1-functional

$$\Phi_\lambda(u) = \int_\Omega |\nabla u|^p - \lambda\, |u|^p$$

defined on the Sobolev space $W_0^{1,p}(\Omega)$ with the usual norm

$$\|u\| = \left(\int_\Omega |\nabla u|^p\right)^{\frac{1}{p}}.$$

When λ is not an eigenvalue, zero is the only critical point of Φ_λ and we may take

$$U = \left\{u \in W_0^{1,p}(\Omega) : \|u\| \leqslant 1\right\}$$

in the definition (1). Since Φ_λ is positive homogeneous, $\Phi_\lambda^0 \cap U$ radially contracts to the origin and $\Phi_\lambda^0 \cap U \backslash \{0\}$ radially deformation retracts onto

$$\Phi_\lambda^0 \cap S = \Psi^\lambda$$

where S is the unit sphere in $W_0^{1,p}(\Omega)$ and

$$\Psi(u) = \frac{1}{\displaystyle\int_\Omega |u|^p}, \qquad u \in S.$$

It follows that

(2)
$$C^q(\Phi_\lambda, 0) \approx \begin{cases} \delta_{q0}\, \mathcal{G}, & \Psi^\lambda = \varnothing \\ \widetilde{H}^{q-1}(\Psi^\lambda), & \Psi^\lambda \neq \varnothing \end{cases}$$

where δ is the Kronecker delta, $\mathcal{G}$ is the coefficient group, and $\widetilde{H}$ denotes reduced cohomology. Note also that the eigenvalues coincide with the critical values of Ψ by the Lagrange multiplier rule.

In the semilinear case $p = 2$, the spectrum $\sigma(-\Delta)$ consists of isolated eigenvalues λ_k, repeated according to their multiplicities, satisfying

$$0 < \lambda_1 < \lambda_2 \leqslant \cdots \to \infty.$$

If $\lambda < \lambda_1 = \inf \Psi$, then $\Psi^\lambda = \varnothing$ and hence

(3)
$$C^q(\Phi_\lambda, 0) \approx \delta_{q0}\, \mathcal{G}$$

by (2). If $\lambda_k < \lambda < \lambda_{k+1}$, then we have the orthogonal decomposition

(4)
$$H_0^1(\Omega) = H^- \oplus H^+, \qquad u = v + w$$

where H^- is the direct sum of the eigenspaces corresponding to $\lambda_1, \ldots, \lambda_k$ and H^+ is its orthogonal complement, and

$$\dim H^- = k$$

is called the Morse index of zero. It is easy to check that

$$\eta(u,t) = \frac{v + (1-t)\,w}{\|v + (1-t)\,w\|}, \qquad (u,t) \in \Psi^\lambda \times [0,1]$$

is a deformation retraction of Ψ^λ onto $H^- \cap S$, so

$$C^q(\Phi_\lambda, 0) \approx \widetilde{H}^{q-1}(H^- \cap S) \approx \delta_{qk}\, \mathcal{G}$$

by (2).

The quasilinear case $p \neq 2$ is far more complicated. Very little is known about the spectrum $\sigma(-\Delta_p)$ itself. The first eigenvalue λ_1 is positive, simple, and has an associated eigenfunction φ_1 that is positive in Ω (see Anane [**9**] and Lindqvist [**68, 69**]). Moreover, λ_1 is isolated in the spectrum, so the second eigenvalue $\lambda_2 = \inf \sigma(-\Delta_p) \cap (\lambda_1, \infty)$ is also well-defined. In the ODE case $n = 1$, where Ω is an interval, the spectrum consists of a sequence of simple eigenvalues $\lambda_k \nearrow \infty$, and the eigenfunction φ_k associated with λ_k has exactly $k-1$ interior zeroes (see, e.g., Drábek [**46**]). In the PDE case $n \geqslant 2$, an increasing and unbounded sequence of eigenvalues can be constructed using a standard minimax scheme involving the Krasnoselskii's genus, but it is not known whether this gives a complete list of the eigenvalues.

If $\lambda < \lambda_1$, then (3) holds as before. It was shown in Dancer and Perera [**40**] that

$$C^q(\Phi_\lambda, 0) \approx \delta_{q1}\, \mathcal{G}$$

if $\lambda_1 < \lambda < \lambda_2$ and that

$$C^q(\Phi_\lambda, 0) = 0, \quad q = 0, 1$$

if $\lambda > \lambda_2$. Thus, the question arises as to whether there is a nontrivial critical group when $\lambda > \lambda_2$. An affirmative answer was given in Perera [**98**] where a new sequence of eigenvalues was constructed using a minimax scheme involving the $\mathbb{Z}_2$-cohomological index of Fadell and Rabinowitz [**49**] as follows.

Let $\mathcal{F}$ denote the class of symmetric subsets of S, let $i(M)$ denote the cohomological index of $M \in \mathcal{F}$, and set

$$\lambda_k := \inf_{\substack{M \in \mathcal{F} \\ i(M) \geqslant k}} \ \sup_{u \in M} \Psi(u).$$

Then $\lambda_k \nearrow \infty$ is a sequence of eigenvalues, and if $\lambda_k < \lambda_{k+1}$, then

(5) $$i(\Psi^{\lambda_k}) = i(S \backslash \Psi_{\lambda_{k+1}}) = k$$

where

$$\Psi^{\lambda_k} = \{u \in S : \Psi(u) \leqslant \lambda_k\}, \qquad \Psi_{\lambda_{k+1}} = \{u \in S : \Psi(u) \geqslant \lambda_{k+1}\}$$

(see Theorem 4.6). Thus, if $\lambda_k < \lambda < \lambda_{k+1}$, then

$$i(\Psi^\lambda) = k$$

by the monotonicity of the index, which implies that

$$\widetilde{H}^{k-1}(\Psi^\lambda) \neq 0$$

(see Proposition 2.14) and hence

$$C^k(\Phi_\lambda, 0) \neq 0 \tag{6}$$

by (2).

The structure provided by this new sequence of eigenvalues is sufficient to adapt many of the standard variational methods for solving semilinear problems to the quasilinear case. In particular, we will construct new linking sets and local linkings that are readily applicable to quasilinear problems. Of course, such constructions cannot be based on linear subspaces since we no longer have eigenspaces to work with. They will instead use nonlinear splittings generated by the sub- and superlevel sets of Ψ that appear in (5), and the indices given there will play a key role in these new topological constructions as we will see next.

Consider the boundary value problem

$$\begin{cases} -\Delta_p\, u = f(x, u) & \text{in } \Omega \\ \quad u = 0 & \text{on } \partial\Omega \end{cases} \tag{7}$$

where the nonlinearity f is a Carathéodory function on $\Omega \times \mathbb{R}$ satisfying the subcritical growth condition

$$|f(x,t)| \leqslant C\left(|t|^{r-1} + 1\right) \quad \forall (x,t) \in \Omega \times \mathbb{R}$$

for some $r \in (1, p^*)$. Here

$$p^* = \begin{cases} \dfrac{np}{n-p}, & p < n \\ \infty, & p \geqslant n \end{cases}$$

is the critical exponent for the Sobolev imbedding $W_0^{1,p}(\Omega) \hookrightarrow L^r(\Omega)$. Weak solutions of this problem coincide with the critical points of the C^1-functional

$$\Phi(u) = \int_\Omega |\nabla u|^p - p\, F(x,u), \quad u \in W_0^{1,p}(\Omega)$$

where

$$F(x,t) = \int_0^t f(x,s)\, ds$$

is the primitive of f.

It is customary to roughly classify problem (7) according to the growth of f as

(i) p-sublinear if

$$\lim_{t \to \pm\infty} \frac{f(x,t)}{|t|^{p-2}\, t} = 0 \quad \forall x \in \Omega,$$

(ii) asymptotically p-linear if

$$0 < \liminf_{t \to \pm\infty} \frac{f(x,t)}{|t|^{p-2}\, t} \leqslant \limsup_{t \to \pm\infty} \frac{f(x,t)}{|t|^{p-2}\, t} < \infty \quad \forall x \in \Omega,$$

(*iii*) p-superlinear if

$$\lim_{t \to \pm\infty} \frac{f(x,t)}{|t|^{p-2}\,t} = \infty \quad \forall x \in \Omega.$$

Consider the asymptotically p-linear case where

$$\lim_{t \to \pm\infty} \frac{f(x,t)}{|t|^{p-2}\,t} = \lambda, \text{ uniformly in } x \in \Omega$$

with $\lambda_k < \lambda < \lambda_{k+1}$, and assume $\lambda \notin \sigma(-\Delta_p)$ to ensure that Φ satisfies the (PS) condition.

In the semilinear case $p = 2$, let

$$A = \big\{v \in H^- : \|v\| = R\big\}, \qquad B = H^+$$

with $H^\pm$ as in (4) and $R > 0$. Then

$$\max \Phi(A) < \inf \Phi(B) \tag{8}$$

if R is sufficiently large, and A cohomologically links B in dimension $k-1$ in the sense that the homomorphism

$$\widetilde{H}^{k-1}(H^1_0(\Omega)\backslash B) \to \widetilde{H}^{k-1}(A)$$

induced by the inclusion is nontrivial. So it follows that problem (7) has a solution u with $C^k(\Phi, u) \neq 0$ (see Proposition 3.25).

We may ask whether this well-known argument can be modified to obtain the same result in the quasilinear case $p \neq 2$ where we no longer have the splitting given in (4). We will give an affirmative answer as follows. Let

$$A = \big\{Ru : u \in \Psi^{\lambda_k}\big\}, \qquad B = \big\{tu : u \in \Psi_{\lambda_{k+1}},\, t \geqslant 0\big\}$$

with $R > 0$. Then (8) still holds if R is sufficiently large, and A cohomologically links B in dimension $k-1$ by (5) and the following theorem proved in Section 3.7, so problem (7) again has a solution u with $C^k(\Phi, u) \neq 0$.

THEOREM 1. *Let A_0 and B_0 be disjoint nonempty closed symmetric subsets of the unit sphere S in a Banach space such that*

$$i(A_0) = i(S\backslash B_0) = k$$

where i denotes the cohomological index, and let

$$A = \big\{Ru : u \in A_0\big\}, \qquad B = \big\{tu : u \in B_0,\, t \geqslant 0\big\}$$

with $R > 0$. Then A cohomologically links B in dimension $k-1$.

Now suppose $f(x, 0) \equiv 0$, so that problem (7) has the trivial solution $u(x) \equiv 0$. Assume that

$$\lim_{t \to 0} \frac{f(x,t)}{|t|^{p-2}\,t} = \lambda, \text{ uniformly in } x \in \Omega, \tag{9}$$

$\lambda_k < \lambda < \lambda_{k+1}$, and the sign condition

$$p\,F(x,t) \geqslant \lambda_{k+1}\,|t|^p \quad \forall (x,t) \in \Omega \times \mathbb{R} \tag{10}$$

holds. In the p-superlinear case it is customary to also assume the following Ambrosetti-Rabinowitz type condition to ensure that Φ satisfies the (PS) condition:

$$0 < \mu F(x,t) \leqslant t f(x,t) \quad \forall x \in \Omega,\ |t| \text{ large} \tag{11}$$

for some $\mu > p$.

In the semilinear case $p = 2$, we can then obtain a nontrivial solution of problem (7) using the well-known saddle point theorem of Rabinowitz as follows. Fix a $w_0 \in H^+ \setminus \{0\}$ and let

$$X = \big\{u = v + s\, w_0 : v \in H^-,\ s \geqslant 0,\ \|u\| \leqslant R\big\},$$

$$A = \big\{v \in H^- : \|v\| \leqslant R\big\} \cup \big\{u \in X : \|u\| = R\big\},$$

$$B = \big\{w \in H^+ : \|w\| = r\big\}$$

with $H^\pm$ as in (4) and $R > r > 0$. Then

$$\max \Phi(A) \leqslant 0 < \inf \Phi(B) \tag{12}$$

if R is sufficiently large and r is sufficiently small, and A homotopically links B with respect to X in the sense that

$$\gamma(X) \cap B \neq \varnothing \quad \forall \gamma \in \Gamma$$

where

$$\Gamma = \big\{\gamma \in C(X, H^1_0(\Omega)) : \gamma|_A = id_A\big\}.$$

So it follows that

$$c := \inf_{\gamma \in \Gamma}\ \sup_{u \in \gamma(X)}\ \Phi(u)$$

is a positive critical level of Φ (see Proposition 3.21).

Again we may ask whether linking sets that would enable us to use this argument in the quasilinear case $p \neq 2$ can be constructed. In Perera and Szulkin [**105**] the following such construction based on the piercing property of the index (see Proposition 2.12) was given. Recall that the cone CA_0 on a topological space A_0 is the quotient space of $A_0 \times [0,1]$ obtained by collapsing $A_0 \times \{1\}$ to a point. We identify $A_0 \times \{0\}$ with A_0 itself. Fix an $h \in C(C\Psi^{\lambda_k}, S)$ such that $h(C\Psi^{\lambda_k})$ is closed and $h|_{\Psi^{\lambda_k}} = id_{\Psi^{\lambda_k}}$, and let

$$X = \big\{tu : u \in h(C\Psi^{\lambda_k}),\ 0 \leqslant t \leqslant R\big\},$$

$$A = \big\{tu : u \in \Psi^{\lambda_k},\ 0 \leqslant t \leqslant R\big\} \cup \big\{u \in X : \|u\| = R\big\},$$

$$B = \big\{ru : u \in \Psi_{\lambda_{k+1}}\big\}$$

with $R > r > 0$. Then (12) still holds if R is sufficiently large and r is sufficiently small, and A homotopically links B with respect to X by (5) and the following theorem proved in Section 3.6, so Φ again has a positive critical level.

THEOREM 2. *Let A_0 and B_0 be disjoint nonempty closed symmetric subsets of the unit sphere S in a Banach space such that*

$$i(A_0) = i(S \backslash B_0) < \infty,$$

$h \in C(CA_0, S)$ be such that $h(CA_0)$ is closed and $h|_{A_0} = id_{A_0}$, and let

$$X = \{tu : u \in h(CA_0),\ 0 \leqslant t \leqslant R\},$$

$$A = \{tu : u \in A_0,\ 0 \leqslant t \leqslant R\} \cup \{u \in X : \|u\| = R\},$$

$$B = \{ru : u \in B_0\}$$

with $R > r > 0$. Then A homotopically links B with respect to X.

The sign condition (10) can be removed by using a comparison of the critical groups of Φ at zero and infinity instead of the above linking argument. First consider the nonresonant case where (9) holds with $\lambda \in (\lambda_k, \lambda_{k+1}) \backslash \sigma(-\Delta_p)$. Then

$$C^q(\Phi, 0) \approx C^q(\Phi_\lambda, 0) \quad \forall q$$

by the homotopy invariance of the critical groups and hence

$$C^k(\Phi, 0) \neq 0 \tag{13}$$

by (6). On the other hand, a simple modification of an argument due to Wang [**132**] shows that

$$C^q(\Phi, \infty) = 0 \quad \forall q$$

when (11) holds (see Example 5.14). So Φ has a nontrivial critical point by the remarks at the beginning of the chapter.

In the p-sublinear case, where Φ is bounded from below, we can use (13) to obtain multiple nontrivial solutions of problem (7). Indeed, the three critical points theorem (see Corollary 3.32) gives two nontrivial critical points of Φ when $k \geqslant 2$.

Note that we do not assume that there are no other eigenvalues in the interval $(\lambda_k, \lambda_{k+1})$, in particular, λ may be an eigenvalue. Our results hold as long as $\lambda_k < \lambda < \lambda_{k+1}$, even if the entire interval $[\lambda_k, \lambda_{k+1}]$ is contained in the spectrum. Thus, eigenvalues that do not belong to the sequence (λ_k) are not that important in this context. In fact, we will see that the cohomological index of sublevel sets changes only when crossing an eigenvalue from this particular sequence.

Now we consider the resonant case where (9) holds with $\lambda \in [\lambda_k, \lambda_{k+1}] \cap \sigma(-\Delta_p)$, and ask whether we still have (13). We will show that this is indeed the case when a suitable sign condition holds near $t = 0$. Write f as

$$f(x,t) = \lambda\, |t|^{p-2}\, t + g(x,t),$$

so that

$$\lim_{t \to 0} \frac{g(x,t)}{|t|^{p-2}\, t} = 0, \quad \text{uniformly in } x \in \Omega,$$

set

$$G(x,t) = \int_0^t g(x,s)\,ds,$$

and assume that either

$$\lambda = \lambda_k, \qquad G(x,t) \geqslant 0 \quad \forall x \in \Omega,\ |t| \text{ small},$$

or

$$\lambda = \lambda_{k+1}, \qquad G(x,t) \leqslant 0 \quad \forall x \in \Omega,\ |t| \text{ small}.$$

In the semilinear case $p = 2$, let

$$A = \big\{v \in H^- : \|v\| \leqslant r\big\}, \qquad B = \big\{w \in H^+ : \|w\| \leqslant r\big\}$$

with $H^{\pm}$ as in (4) and $r > 0$. Then

$$\Phi|_A \leqslant 0 < \Phi|_{B\setminus\{0\}} \tag{14}$$

if r is sufficiently small (see Li and Willem [**67**]), so Φ has a local linking near zero in dimension k and hence (13) holds (see Liu [**71**]).

So we may ask whether the notion of a local linking can be generalized to apply in the quasilinear case $p \neq 2$ as well. We will again give an affirmative answer. Let

$$A = \big\{tu : u \in \Psi^{\lambda_k},\ 0 \leqslant t \leqslant r\big\}, \qquad B = \big\{tu : u \in \Psi_{\lambda_{k+1}},\ 0 \leqslant t \leqslant r\big\}$$

with $r > 0$. Then (14) still holds if r is sufficiently small (see Degiovanni, Lancelotti, and Perera [**42**]), so Φ has a cohomological local splitting near zero in dimension k in the sense of the following definition given in Section 3.11. Hence (13) holds again (see Proposition 3.34).

DEFINITION 3. We say that a C^1-functional Φ defined on a Banach space W has a cohomological local splitting near zero in dimension k if there is an $r > 0$ such that zero is the only critical point of Φ in

$$U = \big\{u \in W : \|u\| \leqslant r\big\}$$

and there are disjoint nonempty closed symmetric subsets A_0 and B_0 of ∂U such that

$$i(A_0) = i(S\setminus B_0) = k$$

and

$$\Phi|_A \leqslant 0 < \Phi|_{B\setminus\{0\}}$$

where

$$A = \big\{tu : u \in A_0,\ 0 \leqslant t \leqslant 1\big\}, \qquad B = \big\{tu : u \in B_0,\ 0 \leqslant t \leqslant 1\big\}.$$

These constructions, which were based on the existence of a sequence of eigenvalues satisfying (5), can be extended to situations involving indefinite eigenvalue problems such as

$$\begin{cases} -\Delta_p\, u = \lambda\, V(x)\, |u|^{p-2}\, u & \text{in } \Omega \\ \quad u = 0 & \text{on } \partial\Omega \end{cases}$$

where the weight function $V \in L^\infty(\Omega)$ changes sign. Here the eigenfunctions are the critical points of the functional

$$\Phi_\lambda(u) = \int_\Omega |\nabla u|^p - \lambda\, V(x)\, |u|^p, \quad u \in W_0^{1,p}(\Omega)$$

and the positive and negative eigenvalues are the critical values of

$$\Psi^\pm(u) = \frac{1}{J(u)}, \quad u \in S^\pm,$$

respectively, where

$$J(u) = \int_\Omega V(x)\, |u|^p$$

and

$$S^\pm = \{u \in S : J(u) \gtrless 0\}.$$

Let $\mathcal{F}^\pm$ denote the class of symmetric subsets of $S^\pm$, respectively, and set

$$\lambda_k^+ := \inf_{\substack{M \in \mathcal{F}^+ \\ i(M) \geqslant k}} \ \sup_{u \in M} \ \Psi^+(u),$$

$$\lambda_k^- := \sup_{\substack{M \in \mathcal{F}^- \\ i(M) \geqslant k}} \ \inf_{u \in M} \ \Psi^-(u).$$

We will show that $\lambda_k^+ \nearrow +\infty$ and $\lambda_k^- \searrow -\infty$ are sequences of positive and negative eigenvalues, respectively, and if $\lambda_k^+ < \lambda_{k+1}^+$ (resp. $\lambda_{k+1}^- < \lambda_k^-$), then

$$i((\Psi^+)^{\lambda_k^+}) = i(S^+ \backslash (\Psi^+)_{\lambda_{k+1}^+}) = k$$

$$(\text{resp. } i((\Psi^-)_{\lambda_k^-}) = i(S^- \backslash (\Psi^-)^{\lambda_{k+1}^-}) = k).$$

In particular, if $\lambda_k^+ < \lambda < \lambda_{k+1}^+$ or $\lambda_{k+1}^- < \lambda < \lambda_k^-$, then

$$C^k(\Phi_\lambda, 0) \neq 0.$$

Finally we will present an extension of our theory to anisotropic p-Laplacian systems of the form

$$(15) \qquad \begin{cases} -\Delta_{p_i} u_i = \dfrac{\partial F}{\partial u_i}(x, u) & \text{in } \Omega \\ \\ u_i = 0 & \text{on } \partial\Omega, \end{cases} \qquad i = 1, \dots, m$$

where each $p_i \in (1, \infty)$, $u = (u_1, \dots, u_m)$, and $F \in C^1(\Omega \times \mathbb{R}^m)$ satisfies the subcritical growth conditions

$$\left|\frac{\partial F}{\partial u_i}(x, u)\right| \leqslant C \left(\sum_{j=1}^m |u_j|^{r_{ij}-1} + 1\right) \quad \forall (x, u) \in \Omega \times \mathbb{R}^m,\ i = 1, \dots, m$$

for some $r_{ij} \in (1, 1 + (1 - 1/p_i^*)\, p_j^*)$. Weak solutions of this system are the critical points of the functional

$$\Phi(u) = I(u) - \int_\Omega F(x, u), \quad u \in W = W_0^{1,p_1}(\Omega) \times \cdots \times W_0^{1,p_m}(\Omega)$$

where

$$I(u) = \sum_{i=1}^m \frac{1}{p_i} \int_\Omega |\nabla u_i|^{p_i}.$$

Unlike in the scalar case, here I is not homogeneous except when $p_1 = \cdots = p_m$. However, it still has the following weaker property. Define a continuous flow on W by

$$\mathbb{R} \times W \to W, \quad (\alpha, u) \mapsto u_\alpha := (|\alpha|^{1/p_1 - 1}\, \alpha\, u_1, \ldots, |\alpha|^{1/p_m - 1}\, \alpha\, u_m).$$

Then

$$I(u_\alpha) = |\alpha|\, I(u) \quad \forall \alpha \in \mathbb{R},\ u \in W.$$

This suggests that the appropriate class of eigenvalue problems to study here are of the form

$$\text{(16)} \qquad \begin{cases} -\Delta_{p_i} u_i = \lambda\, \dfrac{\partial J}{\partial u_i}(x, u) & \text{in } \Omega \\ \\ u_i = 0 & \text{on } \partial\Omega, \end{cases} \qquad i = 1, \ldots, m$$

where $J \in C^1(\Omega \times \mathbb{R}^m)$ satisfies

$$\text{(17)} \qquad J(x, u_\alpha) = |\alpha|\, J(x, u) \quad \forall \alpha \in \mathbb{R},\ (x, u) \in \Omega \times \mathbb{R}^m.$$

For example,

$$J(x, u) = V(x)\, |u_1|^{r_1} \cdots |u_m|^{r_m}$$

where $r_i \in (1, p_i)$ with $r_1/p_1 + \cdots + r_m/p_m = 1$ and $V \in L^\infty(\Omega)$. Note that (17) implies that if u is an eigenvector associated with λ, then so is u_α for any $\alpha \neq 0$.

The eigenfunctions of problem (16) are the critical points of the functional

$$\Phi_\lambda(u) = I(u) - \lambda\, J(u), \quad u \in W$$

where

$$J(u) = \int_\Omega J(x, u).$$

Let

$$\mathcal{M} = \big\{u \in W : I(u) = 1\big\}$$

and suppose that

$$\mathcal{M}^\pm = \big\{u \in \mathcal{M} : J(u) \gtrless 0\big\} \neq \emptyset.$$

Then $\mathcal{M} \subset W \backslash \{0\}$ is a bounded complete symmetric C^1-Finsler manifold radially homeomorphic to the unit sphere in W, $\mathcal{M}^\pm$ are symmetric open submanifolds of $\mathcal{M}$, and the positive and negative eigenvalues are given by the critical values of

$$\Psi^{\pm}(u) = \frac{1}{J(u)}, \quad u \in \mathcal{M}^{\pm},$$

respectively.

Let $\mathcal{F}^{\pm}$ denote the class of symmetric subsets of $\mathcal{M}^{\pm}$, respectively, and set

$$\lambda_k^+ := \inf_{\substack{M \in \mathcal{F}^+ \\ i(M) \geqslant k}} \sup_{u \in M} \Psi^+(u),$$

$$\lambda_k^- := \sup_{\substack{M \in \mathcal{F}^- \\ i(M) \geqslant k}} \inf_{u \in M} \Psi^-(u).$$

We will again show that $\lambda_k^+ \nearrow +\infty$ and $\lambda_k^- \searrow -\infty$ are sequences of positive and negative eigenvalues, respectively, and if $\lambda_k^+ < \lambda_{k+1}^+$ (resp. $\lambda_{k+1}^- < \lambda_k^-$), then

$$i((\Psi^+)^{\lambda_k^+}) = i(\mathcal{M}^+ \backslash (\Psi^+)_{\lambda_{k+1}^+}) = k$$

$$(\text{resp. } i((\Psi^-)_{\lambda_k^-}) = i(\mathcal{M}^- \backslash (\Psi^-)^{\lambda_{k+1}^-}) = k),$$

in particular, if $\lambda_k^+ < \lambda < \lambda_{k+1}^+$ or $\lambda_{k+1}^- < \lambda < \lambda_k^-$, then

$$C^k(\Phi_\lambda, 0) \neq 0.$$

This will allow us to extend our existence and multiplicity theory for a single equation to systems.

For example, suppose $F(x, 0) \equiv 0$, so that the system (15) has the trivial solution $u(x) \equiv 0$. Assume that

$$F(x, u) = \lambda\, J(x, u) + G(x, u)$$

where λ is not an eigenvalue of (16) and

$$|G(x, u)| \leqslant C \sum_{i=1}^{m} |u_i|^{s_i} \quad \forall (x, u) \in \Omega \times \mathbb{R}^m$$

for some $s_i \in (p_i, p_i^*)$. Further assume the following superlinearity condition: there are $\mu_i > p_i$, $i = 1, \ldots, m$ such that $\displaystyle\sum_{i=1}^{m} \left(\frac{1}{p_i} - \frac{1}{\mu_i}\right) u_i \frac{\partial F}{\partial u_i}$ is bounded from below and

$$0 < F(x, u) \leqslant \sum_{i=1}^{m} \frac{u_i}{\mu_i} \frac{\partial F}{\partial u_i}(x, u) \quad \forall x \in \Omega,\ |u| \text{ large.}$$

We will obtain a nontrivial solution of (15) under these assumptions in Sections 10.2 and 10.3.

All this machinery can be adapted to many other p-Laplacian like operators as well. Therefore we will develop our theory in an abstract operator setting that includes many of them as special cases.

CHAPTER 0

Morse Theory and Variational Problems

In this preliminary chapter we give a brief survey of Morse theoretic methods used in variational problems. General references are Milnor [**84**], Mawhin and Willem [**81**], Chang [**28**], and Benci [**19**].

We consider a real-valued function Φ defined on a real Banach space $(W, \|\cdot\|)$. We say that Φ is Fréchet differentiable at $u \in W$ if there is an element $\Phi'(u)$ of the dual space $(W^*, \|\cdot\|^*)$, called the Fréchet derivative of Φ at u, such that

$$\Phi(u + v) = \Phi(u) + \big\langle \Phi'(u), v \big\rangle + \mathrm{o}(\|v\|) \text{ as } v \to 0 \text{ in } W,$$

where $\langle \cdot, \cdot \rangle$ is the duality pairing. The functional Φ is continuously Fréchet differentiable on W, or belongs to the class $C^1(W, \mathbb{R})$, if Φ' is defined everywhere and the map $W \to W^*$, $u \mapsto \Phi'(u)$ is continuous. We assume that $\Phi \in C^1(W, \mathbb{R})$ for the rest of the chapter. We say that u is a critical point of Φ if $\Phi'(u) = 0$. A real number $c \in \Phi(W)$ is a critical value of Φ if there is a critical point u with $\Phi(u) = c$, otherwise it is a regular value. We use the notations

$$\Phi_a = \big\{u \in W : \Phi(u) \geqslant a\big\}, \quad \Phi^b = \big\{u \in W : \Phi(u) \leqslant b\big\}, \quad \Phi_a^b = \Phi_a \cap \Phi^b,$$

$$K = \big\{u \in W : \Phi'(u) = 0\big\}, \quad \widetilde{W} = W \backslash K, \quad K_a^b = K \cap \Phi_a^b, \quad K^c = K_c^c$$

for the various superlevel, sublevel, critical, and regular sets of Φ.

We begin by recalling the compactness condition of Palais and Smale and its weaker variant given by Cerami in Section 0.1. Then we state the first and second deformation lemmas under the Cerami's condition in Section 0.2. In Section 0.3 we define the critical groups of an isolated critical point and summarize the basic results of Morse theory. These include the Morse inequalities, Morse lemma and its generalization splitting lemma, shifting theorem of Gromoll and Meyer, and the handle body theorem. Next we discuss the minimax principle in Section 0.4. Section 0.5 contains a discussion of homotopical linking, pairs of critical points with nontrivial critical groups produced by homological linking, and nonstandard geometries without a finite dimensional closed loop. We recall the notion of local linking and an alternative for a critical point produced by a local linking in Section 0.6. We conclude with a result on nontrivial critical groups associated with the p-Laplacian in Section 0.7.

0.1. Compactness Conditions

It is usually necessary to assume some sort of a "compactness condition" when seeking critical points of a functional. The following condition was originally introduced by Palais and Smale [**91**]: Φ satisfies the Palais-Smale compactness condition at the level c, or $(\mathrm{PS})_c$ for short, if every sequence $(u_j) \subset W$ such that

$$\Phi(u_j) \to c, \quad \Phi'(u_j) \to 0,$$

called a $(\mathrm{PS})_c$ sequence, has a convergent subsequence; Φ satisfies (PS) if it satisfies $(\mathrm{PS})_c$ for every $c \in \mathbb{R}$, or equivalently, if every sequence such that $\Phi(u_j)$ is bounded and $\Phi'(u_j) \to 0$, called a (PS) sequence, has a convergent subsequence.

The following weaker version was introduced by Cerami [**25**]: Φ satisfies the Cerami condition at the level c, or $(\mathrm{C})_c$ for short, if every sequence such that

$$\Phi(u_j) \to c, \quad \big(1 + \|u_j\|\big)\, \Phi'(u_j) \to 0,$$

called a $(\mathrm{C})_c$ sequence, has a convergent subsequence; Φ satisfies (C) if it satisfies $(\mathrm{C})_c$ for every c, or equivalently, if every sequence such that $\Phi(u_j)$ is bounded and $\big(1 + \|u_j\|\big)\, \Phi'(u_j) \to 0$, called a (C) sequence, has a convergent subsequence. This condition is weaker since a $(\mathrm{C})_c$ (resp. (C)) sequence is clearly a $(\mathrm{PS})_c$ (resp. (PS)) sequence also.

The limit of a $(\mathrm{PS})_c$ (resp. (PS)) sequence is in K^c (resp. K) since Φ and Φ' are continuous. Since any sequence in K^c is a $(\mathrm{C})_c$ sequence, it follows that K^c is a compact set when $(\mathrm{C})_c$ holds.

0.2. Deformation Lemmas

An essential tool for locating critical points is the deformation lemmas, which allow to lower sublevel sets of a functional, away from its critical set. The main ingredient in their proofs is a suitable negative pseudo-gradient flow, a notion due to Palais [**93**]: a pseudo-gradient vector field for Φ on $\widetilde{W}$ is a locally Lipschitz continuous mapping $V : \widetilde{W} \to W$ satisfying

$$\|V(u)\| \leqslant \big\|\Phi'(u)\big\|^*, \;\; 2\,\big\langle \Phi'(u), V(u) \big\rangle \geqslant \big(\big\|\Phi'(u)\big\|^*\big)^2 \quad \forall u \in \widetilde{W}.$$

Such a vector field exists, and may be chosen to be odd when Φ is even.

The first deformation lemma provides a local deformation near a (possibly critical) level set of a functional.

Lemma 0.1 (First Deformation Lemma). *If $c \in \mathbb{R}$, C is a bounded set containing K^c, $\delta, k > 0$, and Φ satisfies $(\mathrm{C})_c$, then there are an $\varepsilon_0 > 0$ and, for each $\varepsilon \in (0, \varepsilon_0)$, a map $\eta \in C([0,1] \times W, W)$ satisfying*

(i) $\eta(0, \cdot) = id_W$,
(ii) *$\eta(t, \cdot)$ is a homeomorphism of W for all $t \in [0,1]$,*
(iii) *$\eta(t, \cdot)$ is the identity outside $A = \Phi_{c-2\varepsilon}^{c+2\varepsilon} \backslash N_{\delta/3}(C)$ for all $t \in [0,1]$,*
(iv) $\|\eta(t,u) - u\| \leqslant \big(1 + \|u\|\big)\, \delta/k \quad \forall (t,u) \in [0,1] \times W$,
(v) *$\Phi(\eta(\cdot, u))$ is nonincreasing for all $u \in W$,*

(*vi*) $\eta(1, \Phi^{c+\varepsilon} \backslash N_\delta(C)) \subset \Phi^{c-\varepsilon}$.

When Φ *is even and* C *is symmetric,* η *may be chosen so that* $\eta(t, \cdot)$ *is odd for all* $t \in [0,1]$.

First deformation lemma under the $(\mathrm{PS})_c$ condition is due to Palais [**92**]; see also Rabinowitz [**108**]. The proof under the $(\mathrm{C})_c$ condition was given by Cerami [**25**] and Bartolo, Benci, and Fortunato [**13**]. The particular version given here will be proved in Section 3.2.

The second deformation lemma implies that the homotopy type of sublevel sets can change only when crossing a critical level.

Lemma 0.2 (Second Deformation Lemma). *If* $-\infty < a < b \leqslant +\infty$ *and* Φ *has only a finite number of critical points at the level* a*, has no critical values in* (a, b)*, and satisfies* $(\mathrm{C})_c$ *for all* $c \in [a, b] \cap \mathbb{R}$*, then* Φ^a *is a deformation retract of* $\Phi^b \backslash K^b$*, i.e., there is a map* $\eta \in C([0,1] \times (\Phi^b \backslash K^b), \Phi^b \backslash K^b)$*, called a deformation retraction of* $\Phi^b \backslash K^b$ *onto* Φ^a*, satisfying*

(*i*) $\eta(0, \cdot) = id_{\Phi^b \backslash K^b}$,
(*ii*) $\eta(t, \cdot)|_{\Phi^a} = id_{\Phi^a} \quad \forall t \in [0,1]$,
(*iii*) $\eta(1, \Phi^b \backslash K^b) = \Phi^a$.

Second deformation lemma under the $(\mathrm{PS})_c$ condition is due to Rothe [**117**], Chang [**27**], and Wang [**131**]. The proof under the $(\mathrm{C})_c$ condition can be found in Bartsch and Li [**14**], Perera and Schechter [**104**], and in Section 3.2.

0.3. Critical Groups

In Morse theory the local behavior of Φ near an isolated critical point u is described by the sequence of critical groups

$$C_q(\Phi, u) = H_q(\Phi^c \cap U, \Phi^c \cap U \backslash \{u\}), \quad q \geqslant 0 \tag{0.1}$$

where $c = \Phi(u)$ is the corresponding critical value, U is a neighborhood of u containing no other critical points, and H denotes singular homology. They are independent of the choice of U by the excision property.

For example, if u is a local minimizer, $C_q(\Phi, u) = \delta_{q0}\, \mathcal{G}$ where δ is the Kronecker delta and $\mathcal{G}$ is the coefficient group. A critical point u with $C_1(\Phi, u) \neq 0$ is called a mountain pass point.

Let $-\infty < a < b \leqslant +\infty$ be regular values and assume that Φ has only isolated critical values $c_1 < c_2 < \cdots$ in (a, b), with a finite number of critical points at each level, and satisfies $(\mathrm{PS})_c$ for all $c \in [a, b] \cap \mathbb{R}$. Then the Morse type numbers of Φ with respect to the interval (a, b) are defined by

$$M_q(a, b) = \sum_i \operatorname{rank} H_q(\Phi^{a_{i+1}}, \Phi^{a_i}), \quad q \geqslant 0$$

where $a = a_1 < c_1 < a_2 < c_2 < \cdots$. They are independent of the a_i by the second deformation lemma, and are related to the critical groups by

$$M_q(a,b) = \sum_{u \in K_a^b} \operatorname{rank} C_q(\Phi, u).$$

Writing $\beta_j(a,b) = \operatorname{rank} H_j(\Phi^b, \Phi^a)$, we have

Theorem 0.3 (Morse Inequalities). *If there is only a finite number of critical points in Φ_a^b, then*

$$\sum_{j=0}^{q} (-1)^{q-j} M_j \geqslant \sum_{j=0}^{q} (-1)^{q-j} \beta_j, \quad q \geqslant 0,$$

and

$$\sum_{j=0}^{\infty} (-1)^j M_j = \sum_{j=0}^{\infty} (-1)^j \beta_j$$

when the series converge.

Critical groups are invariant under homotopies that preserve the isolatedness of the critical point; see Rothe [**116**], Chang and Ghoussoub [**26**], and Corvellec and Hantoute [**33**].

Theorem 0.4. *If Φ_t, $t \in [0,1]$ is a family of C^1-functionals on W satisfying* (PS)*, u is a critical point of each Φ_t, and there is a closed neighborhood U such that*

(*i*) *U contains no other critical points of Φ_t,*
(*ii*) *the map $[0,1] \to C^1(U, \mathbb{R})$, $t \mapsto \Phi_t$ is continuous,*

then $C_*(\Phi_t, u)$ are independent of t.

When the critical values are bounded from below and Φ satisfies (C), the global behavior of Φ can be described by the critical groups at infinity introduced by Bartsch and Li [**14**]

$$C_q(\Phi, \infty) = H_q(W, \Phi^a), \quad q \geqslant 0$$

where a is less than all critical values. They are independent of a by the second deformation lemma and the homotopy invariance of the homology groups.

For example, if Φ is bounded from below, $C_q(\Phi, \infty) = \delta_{q0}\, \mathcal{G}$. If Φ is unbounded from below, $C_q(\Phi, \infty) = \widetilde{H}_{q-1}(\Phi^a)$ where $\widetilde{H}$ denotes the reduced groups.

Theorem 0.5. *If $C_k(\Phi, \infty) \neq 0$ and Φ has only a finite number of critical points and satisfies* (C)*, then Φ has a critical point u with $C_k(\Phi, u) \neq 0$.*

The second deformation lemma implies that $C_q(\Phi, \infty) = C_q(\Phi, 0)$ if $u = 0$ is the only critical point of Φ, so Φ has a nontrivial critical point if $C_q(\Phi, 0) \neq C_q(\Phi, \infty)$ for some q.

Now suppose that W is a Hilbert space $(H, (\cdot,\cdot))$ and $\Phi \in C^2(H, \mathbb{R})$. Then the Hessian $A = \Phi''(u)$ is a self-adjoint operator on H for each u. When u is a critical point the dimension of the negative space of A is called the Morse index of u and is denoted by $m(u)$, and $m^*(u) = m(u) + \dim \ker A$ is called the large Morse index. We say that u is nondegenerate if A is invertible. The Morse lemma describes the local behavior of the functional near a nondegenerate critical point.

Lemma 0.6 (Morse Lemma). *If u is a nondegenerate critical point of Φ, then there is a local diffeomorphism ξ from a neighborhood U of u into H with $\xi(u) = 0$ such that*

$$\Phi(\xi^{-1}(v)) = \Phi(u) + \frac{1}{2}\langle Av, v\rangle, \quad v \in \xi(U).$$

Morse lemma in $\mathbb{R}^n$ was proved by Morse [**85**]. Palais [**92**], Schwartz [**123**], and Nirenberg [**88**] extended it to Hilbert spaces when Φ is C^3. Proof in the C^2 case is due to Kuiper [**62**] and Cambini [**23**].

A direct consequence of the Morse lemma is

Theorem 0.7. *If u is a nondegenerate critical point of Φ, then*

$$C_q(\Phi, u) = \delta_{qm(u)}\, \mathcal{G}.$$

The handle body theorem describes the change in topology as the level sets pass through a critical level on which there are only nondegenerate critical points.

Theorem 0.8 (Handle Body Theorem). *If c is an isolated critical value of Φ for which there are only a finite number of nondegenerate critical points u_i, $i = 1, \ldots, k$, with Morse indices $m_i = m(u_i)$, and Φ satisfies* (PS), *then there are an $\varepsilon > 0$ and homeomorphisms φ_i from the unit disk D^{m_i} in $\mathbb{R}^{m_i}$ into H such that*

$$\Phi^{c-\varepsilon} \cap \varphi_i(D^{m_i}) = \Phi^{-1}(c-\varepsilon) \cap \varphi_i(D^{m_i}) = \varphi_i(\partial D^{m_i})$$

and $\Phi^{c-\varepsilon} \cup \bigcup_{i=1}^k \varphi_i(D^{m_i})$ is a deformation retract of $\Phi^{c+\varepsilon}$.

The references for Theorems 0.3, 0.7, and 0.8 are Morse [**85**], Pitcher [**106**], Milnor [**84**], Rothe [**114, 115, 117**], Palais [**92**], Palais and Smale [**91**], Smale [**124**], Marino and Prodi [**79**], Schwartz [**123**], Mawhin and Willem [**81**], and Chang [**28**].

The splitting lemma generalizes the Morse lemma to degenerate critical points. Assume that the origin is an isolated degenerate critical point of Φ and 0 is an isolated point of the spectrum of $A = \Phi''(0)$. Let $N = \ker A$ and write $H = N \oplus N^\perp$, $u = v + w$.

Lemma 0.9 (Splitting Lemma). *There are a ball $B \subset H$ centered at the origin, a local homeomorphism ξ from B into H with $\xi(0) = 0$, and a map $\eta \in C^1(B \cap N, N^\perp)$ such that*

$$\Phi(\xi(u)) = \frac{1}{2}\langle Aw, w\rangle + \Phi(v + \eta(v)), \quad u \in B.$$

Splitting lemma when A is a compact perturbation of the identity was proved by Gromoll and Meyer [**54**] for $\Phi \in C^3$ and by Hofer [**56**] in the C^2 case. Mawhin and Willem [**80, 81**] extended it to the case where A is a Fredholm operator of index zero. The general version given here is due to Chang [**28**].

A consequence of the splitting lemma is

Theorem 0.10 (Shifting Theorem). *We have*

$$C_q(\Phi, 0) = C_{q-m(0)}(\Phi|_{\mathcal{N}}, 0) \quad \forall q$$

where $\mathcal{N} = \xi(B \cap N)$ *is the degenerate submanifold of* Φ *at* 0.

Shifting theorem is due to Gromoll and Meyer [**54**]; see also Mawhin and Willem [**81**] and Chang [**28**].

Since $\dim \mathcal{N} = m^*(0) - m(0)$, shifting theorem gives us the following Morse index estimates when there is a nontrivial critical group.

Corollary 0.11. *If* $C_k(\Phi, 0) \neq 0$, *then*

$$m(0) \leqslant k \leqslant m^*(0).$$

It also enables us to compute the critical groups of a mountain pass point of nullity at most one.

Theorem 0.12. *If* u *is a mountain pass point of* Φ *and* $\dim \ker \Phi''(u) \leqslant 1$, *then*

$$C_q(\Phi, u) = \delta_{q1}\, \mathcal{G}.$$

This result is due to Ambrosetti [**5, 6**] in the nondegenerate case and to Hofer [**56**] in the general case.

Shifting theorem also implies that all critical groups of a critical point with infinite Morse index are trivial, so the above theory is not suitable for studying strongly indefinite functionals. An infinite dimensional Morse theory particularly well suited to deal with such functionals was developed by Szulkin [**128**]; see also Kryszewski and Szulkin [**61**].

The following important perturbation result is due to Marino and Prodi [**78**]; see also Solimini [**125**].

Theorem 0.13. *If some critical value of* Φ *has only a finite number of critical points* u_i, $i = 1, \dots, k$ *and* $\Phi''(u_i)$ *are Fredholm operators, then for any sufficiently small* $\varepsilon > 0$ *there is a* C^2*-functional* Φ_ε *on* H *such that*

(*i*) $\|\Phi_\varepsilon - \Phi\|_{C^2(H)} \leqslant \varepsilon$,

(*ii*) $\Phi_\varepsilon = \Phi$ *in* $H \setminus \bigcup_{i=1}^k B_\varepsilon(u_j)$,

(*iii*) Φ_ε *has only nondegenerate critical points in* $B_\varepsilon(u_j)$ *and their Morse indices are in* $[m(u_i), m^*(u_i)]$,

(*iv*) Φ *satisfies* (PS) $\implies$ Φ_ε *satisfies* (PS).

0.4. Minimax Principle

Minimax principle originated in the work of Ljusternik and Schnirelmann [**76**] and is a useful tool for finding critical points of a functional. Note that the first deformation lemma implies that if c is a regular value and Φ satisfies $(\mathrm{C})_c$, then the family $\mathcal{D}_{c,\varepsilon}$ of maps $\eta \in C([0,1] \times W, W)$ satisfying

(*i*) $\eta(0,\cdot) = id_W$,
(*ii*) $\eta(t,\cdot)$ is a homeomorphism of W for all $t \in [0,1]$,
(*iii*) $\eta(t,\cdot)$ is the identity outside $\Phi_{c-2\varepsilon}^{c+2\varepsilon}$ for all $t \in [0,1]$,
(*iv*) $\Phi(\eta(\cdot,u))$ is nonincreasing for all $u \in W$,
(*v*) $\eta(1,\Phi^{c+\varepsilon}) \subset \Phi^{c-\varepsilon}$

is nonempty for all sufficiently small $\varepsilon > 0$. We say that a family $\mathcal{F}$ of subsets of W is invariant under $\mathcal{D}_{c,\varepsilon}$ if

$$M \in \mathcal{F},\ \eta \in \mathcal{D}_{c,\varepsilon} \implies \eta(1,M) \in \mathcal{F}.$$

Theorem 0.14 (Minimax Principle). *If $\mathcal{F}$ is a family of subsets of W,*

$$c := \inf_{M \in \mathcal{F}} \sup_{u \in M} \Phi(u)$$

is finite, $\mathcal{F}$ is invariant under $\mathcal{D}_{c,\varepsilon}$ for all sufficiently small $\varepsilon > 0$, and Φ satisfies $(\mathrm{C})_c$, then c is a critical value of Φ.

We say that a family Γ of continuous maps γ from a topological space X into W is invariant under $\mathcal{D}_{c,\varepsilon}$ if

$$\gamma \in \Gamma,\ \eta \in \mathcal{D}_{c,\varepsilon} \implies \eta(1,\cdot) \circ \gamma \in \Gamma.$$

Minimax principle is often applied in the following form, which follows by taking $\mathcal{F} = \{\gamma(X) : \gamma \in \Gamma\}$ in Theorem 0.14.

Theorem 0.15. *If Γ is a family of continuous maps γ from a topological space X into W,*

$$c := \inf_{\gamma \in \Gamma} \sup_{u \in \gamma(X)} \Phi(u)$$

is finite, Γ is invariant under $\mathcal{D}_{c,\varepsilon}$ for all sufficiently small $\varepsilon > 0$, and Φ satisfies $(\mathrm{C})_c$, then c is a critical value of Φ.

Some references for Theorems 0.14 and 0.15 are Palais [**93**], Nirenberg [**89**], Rabinowitz [**111**], and Ghoussoub [**53**].

Minimax methods were introduced in Morse theory by Marino and Prodi [**79**]. The following result is due to Liu [**71**].

Theorem 0.16. *If $\sigma \in H_k(\Phi^b, \Phi^a)$ is a nontrivial singular homology class where $-\infty < a < b \leqslant +\infty$ are regular values,*

$$c := \inf_{z \in \sigma} \sup_{u \in |z|} \Phi(u)$$

where $|z|$ denotes the support of the singular chain z, Φ satisfies $(\mathrm{C})_c$, and K^c is a finite set, then there is a $u \in K^c$ with $C_k(\Phi,u) \neq 0$.

0.5. Linking

The notion of homotopical linking is useful for obtaining critical points via the minimax principle.

Definition 0.17. Let A be a closed proper subset of a topological space X, $g \in C(A, W)$ such that $g(A)$ is closed, B a nonempty closed subset of W such that $\operatorname{dist}(g(A), B) > 0$, and

$$\Gamma = \{\gamma \in C(X, W) : \gamma(X) \text{ is closed}, \ \gamma|_A = g\}.$$

We say that (A, g) homotopically links B with respect to X if

$$\gamma(X) \cap B \neq \varnothing \quad \forall \gamma \in \Gamma.$$

When $g : A \subset W$ is the inclusion and $X = \{tu : u \in A, \ t \in [0, 1]\}$, we simply say that A homotopically links B.

Some standard examples of homotopical linking are the following.

Example 0.18. If $u_0 \in W$, U is a bounded neighborhood of u_0, and $u_1 \notin \overline{U}$, then $A = \{u_0, u_1\}$ homotopically links $B = \partial U$.

Example 0.19. If $W = W_1 \oplus W_2$, $u = u_1 + u_2$ is a direct sum decomposition of W with W_1 nontrivial and finite dimensional, then $A = \{u_1 \in W_1 : \|u_1\| = R\}$ homotopically links $B = W_2$ for any $R > 0$.

Example 0.20. If $W = W_1 \oplus W_2$, $u = u_1 + u_2$ is a direct sum decomposition with W_1 finite dimensional and $v \in W_2$ with $\|v\| = 1$, then $A = \{u_1 \in W_1 : \|u_1\| \leqslant R\} \cup \{u = u_1 + tv : u_1 \in W_1, \ t \geqslant 0, \ \|u\| = R\}$ homotopically links $B = \{u_2 \in W_2 : \|u_2\| = r\}$ for any $0 < r < R$.

Theorem 0.21. *If (A, g) homotopically links B with respect to X,*

$$c := \inf_{\gamma \in \Gamma} \sup_{u \in \gamma(X)} \Phi(u)$$

is finite, $a := \sup \Phi(g(A)) \leqslant \inf \Phi(B) =: b$, and Φ satisfies $(\mathrm{C})_c$, then $c \geqslant b$ and is a critical value of Φ. If $c = b$, then Φ has a critical point with critical value c on B.

Many authors have contributed to this result. The special cases that correspond to Examples 0.18, 0.19, and 0.20 are the well-known mountain pass lemma of Ambrosetti and Rabinowitz [**7**] and the saddle point and linking theorems of Rabinowitz [**110, 109**], respectively. See also Ahmad, Lazer, and Paul [**3**], Castro and Lazer [**24**], Benci and Rabinowitz [**20**], Ni [**87**], Chang [**28**], Qi [**107**], and Ghoussoub [**52**]. The version given here can be found in Section 3.6.

Morse index estimates for a critical point produced by a homotopical linking have been obtained by Lazer and Solimini [**65**], Solimini [**125**], Ghoussoub [**52**], Ramos and Sanchez [**112**], and others. However, the notion of homological linking introduced by Benci [**17, 18**] and Liu [**71**] is better suited for obtaining critical points with nontrivial critical groups.

Definition 0.22. Let A and B be disjoint nonempty subsets of W. We say that A homologically links B in dimension k if the inclusion $i : A \subset W \backslash B$ induces a nontrivial homomorphism

$$i_* : \widetilde{H}_k(A) \to \widetilde{H}_k(W \backslash B).$$

In Examples 0.18, 0.19, and 0.20, A homologically links B in dimensions 0, $\dim W_1 - 1$, and $\dim W_1$, respectively.

Theorem 0.23. *If A homologically links B in dimension k, $\Phi|_A \leqslant a < \Phi|_B$ where a is a regular value, and Φ has only a finite number of critical points in Φ_a and satisfies $(\mathrm{C})_c$ for all $c \geqslant a$, then Φ has a critical point u_1 with*

$$\Phi(u_1) > a, \quad C_{k+1}(\Phi, u_1) \neq 0.$$

This follows from Theorem 0.16. Indeed, since the composition $\widetilde{H}_k(A) \to \widetilde{H}_k(\Phi^a) \to \widetilde{H}_k(W \backslash B)$ induced by the inclusions $A \subset \Phi^a \subset W \backslash B$ is i_*, $\widetilde{H}_k(\Phi^a)$ is nontrivial, and since W is contractible, it then follows from the exact sequence of the pair (W, Φ^a) that $H_{k+1}(W, \Phi^a)$ is nontrivial.

Note that when W_1 is infinite dimensional in Examples 0.19 and 0.20 the set A is contractible and therefore does not link B homotopically or homologically. Schechter and Tintarev [**122**] introduced yet another notion of linking according to which A links B in those examples as long as W_1 or W_2 is finite dimensional; see also Ribarska, Tsachev, and Krastanov [**113**] and Schechter [**120, 121**]. Moreover, according to their definition of linking, if A and B are disjoint closed bounded subsets of W such that A links B and $W \backslash A$ is connected, then B links A. If, in addition, $a := \sup \Phi(A) \leqslant \inf \Phi(B) =: b$ and Φ is bounded on bounded sets and satisfies (C), this then yields a pair of critical points u_1 and u_2 with $\Phi(u_1) \geqslant b \geqslant a \geqslant \Phi(u_2)$. The following analogous result for homological linking was obtained in Perera [**94**], where it was shown that the second critical point also has a nontrivial critical group. We assume that Φ has only a finite number of critical points and satisfies (C) for the rest of this section.

Theorem 0.24. *If A homologically links B in dimension k and B is bounded, $\Phi|_A \leqslant a < \Phi|_B$ where a is a regular value, and Φ is bounded from below on bounded sets, then Φ has two critical points u_1 and u_2 with*

$$\Phi(u_1) > a > \Phi(u_2), \quad C_{k+1}(\Phi, u_1) \neq 0,\ C_k(\Phi, u_2) \neq 0.$$

Corollary 0.25. *Let $W = W_1 \oplus W_2$, $u = u_1 + u_2$ be a direct sum decomposition with $\dim W_1 = k < \infty$. If $\Phi \leqslant a$ on $\{u_1 \in W_1 : \|u_1\| \leqslant R\} \cup \{u = u_1 + tv : u_1 \in W_1,\ t \geqslant 0,\ \|u\| = R\}$ for some $R > 0$ and $v \in W_2$ with $\|v\| = 1$, $\Phi > a$ on $\{u_2 \in W_2 : \|u_2\| = r\}$ for some $0 < r < R$, where a is a regular value, and Φ is bounded from below on bounded sets, then Φ has two critical points u_1 and u_2 with*

$$\Phi(u_1) > a > \Phi(u_2), \quad C_{k+1}(\Phi, u_1) \neq 0,\ C_k(\Phi, u_2) \neq 0.$$

It was also shown in Perera [**94**] that the assumptions that B is bounded and Φ is bounded from below on bounded sets can be relaxed as follows; see also Schechter [**118**].

Theorem 0.26. *If A homologically links B in dimension k, $\Phi|_A \leqslant a < \Phi|_B$ where a is a regular value, and Φ is bounded from below on a set $C \supset B$ such that the inclusion-induced homomorphism $\widetilde{H}_k(W \backslash C) \to \widetilde{H}_k(W \backslash B)$ is trivial, then Φ has two critical points u_1 and u_2 with*

$$\Phi(u_1) > a > \Phi(u_2), \quad C_{k+1}(\Phi, u_1) \neq 0,\, C_k(\Phi, u_2) \neq 0.$$

Corollary 0.27. *Let $W = W_1 \bigoplus W_2$, $u = u_1 + u_2$ be a direct sum decomposition with* $\dim W_1 = k < \infty$. *If $\Phi \leqslant a$ on $\{u_1 \in W_1 : \|u_1\| = R\}$ for some $R > 0$, $\Phi > a$ on W_2, where a is a regular value, and Φ is bounded from below on $\{tv + u_2 : t \geqslant 0,\, u_2 \in W_2\}$ for some $v \in W_1 \backslash \{0\}$, then Φ has two critical points u_1 and u_2 with*

$$\Phi(u_1) > a > \Phi(u_2), \quad C_k(\Phi, u_1) \neq 0,\, C_{k-1}(\Phi, u_2) \neq 0.$$

The following theorem of Perera and Schechter [**103**] gives a critical point with a nontrivial critical group in a saddle point theorem with nonstandard geometrical assumptions that do not involve a finite dimensional closed loop; see also Perera and Schechter [**102**] and Lancelotti [**63**].

Theorem 0.28. *Let $W = W_1 \bigoplus W_2$, $u = u_1 + u_2$ be a direct sum decomposition with* $\dim W_1 = k < \infty$. *If Φ is bounded from above on W_1 and from below on W_2, then Φ has a critical point u_1 with*

$$\inf \Phi(W_2) \leqslant \Phi(u_1) \leqslant \sup \Phi(W_1), \quad C_k(\Phi, u_1) \neq 0.$$

0.6. Local Linking

In many applications Φ has the trivial critical point $u = 0$ and we are interested in finding others. The notion of local linking was introduced by Li and Liu [**72, 66**], who used it to obtain nontrivial critical points under various assumptions on the behavior of Φ at infinity; see also Brezis and Nirenberg [**21**] and Li and Willem [**67**].

Definition 0.29. Assume that the origin is a critical point of Φ with $\Phi(0) = 0$. We say that Φ has a local linking near the origin if there is a direct sum decomposition $W = W_1 \bigoplus W_2$, $u = u_1 + u_2$ with W_1 finite dimensional such that

$$\begin{cases} \Phi(u_1) \leqslant 0, & u_1 \in W_1,\, \|u_1\| \leqslant r \\ \\ \Phi(u_2) > 0, & u_2 \in W_2,\, 0 < \|u_2\| \leqslant r \end{cases}$$

for sufficiently small $r > 0$.

Liu [**71**] showed that this yields a nontrivial critical group at the origin.

Theorem 0.30. *If Φ has a local linking near the origin with* $\dim W_1 = k$ *and the origin is an isolated critical point, then $C_k(\Phi, 0) \neq 0$.*

The following alternative obtained in Perera [**95**] gives a nontrivial critical point with a nontrivial critical group produced by a local linking.

Theorem 0.31. *If* Φ *has a local linking near the origin with* $\dim W_1 = k$, $H_k(\Phi^b, \Phi^a) = 0$ *where* $-\infty < a < 0 < b \leqslant +\infty$ *are regular values, and* Φ *has only a finite number of critical points in* Φ_a^b *and satisfies* $(\mathrm{C})_c$ *for all* $c \in [a, b] \cap \mathbb{R}$, *then* Φ *has a critical point* $u_1 \neq 0$ *with either*

$$a < \Phi(u_1) < 0, \quad C_{k-1}(\Phi, u_1) \neq 0$$

or

$$0 < \Phi(u_1) < b, \quad C_{k+1}(\Phi, u_1) \neq 0.$$

When Φ is bounded from below, taking $a < \inf \Phi(W)$ and $b = +\infty$ gives the following three critical points theorem; see also Krasnosel'skii [**60**], Chang [**27**], Li and Liu [**72**], and Liu [**71**].

Corollary 0.32. *If* Φ *has a local linking near the origin with* $\dim W_1 = k \geqslant 2$, *is bounded from below, has only a finite number of critical points, and satisfies* (C), *then* Φ *has a global minimizer* $u_0 \neq 0$ *with*

$$\Phi(u_0) < 0, \quad C_q(\Phi, u_0) = \delta_{q0}\, \mathcal{G}$$

and a critical point $u_1 \neq 0, u_0$ *with either*

$$\Phi(u_1) < 0, \quad C_{k-1}(\Phi, u_1) \neq 0$$

or

$$\Phi(u_1) > 0, \quad C_{k+1}(\Phi, u_1) \neq 0.$$

Theorems 0.30 and 0.31 and Corollary 0.32 also hold under the more general notion of homological local linking introduced in Perera [**96**].

0.7. p-Laplacian

The p-Laplacian operator

$$\Delta_p\, u = \operatorname{div}\left(|\nabla u|^{p-2}\, \nabla u\right), \quad p \in (1, \infty)$$

arises in non-Newtonian fluid flows, turbulent filtration in porous media, plasticity theory, rheology, glacelogy, and in many other application areas; see, e.g., Esteban and Vázquez [**48**] and Padial, Takáč, and Tello [**90**]. Problems involving the p-Laplacian have been studied extensively in the literature during the last fifty years. In this section we present a result on nontrivial critical groups associated with the p-Laplacian obtained in Perera [**98**]; see also Dancer and Perera [**40**].

Consider the nonlinear eigenvalue problem

$$(0.2) \qquad \begin{cases} -\Delta_p\, u = \lambda\, |u|^{p-2}\, u & \text{in } \Omega \\ u = 0 & \text{on } \partial\Omega \end{cases}$$

where Ω is a bounded domain in $\mathbb{R}^n$, $n \geqslant 1$ and $\Delta_p\, u = \operatorname{div}\big(|\nabla u|^{p-2}\, \nabla u\big)$ is the p-Laplacian of u, $p \in (1, \infty)$. It is known that the first eigenvalue λ_1 is positive, simple, has an associated eigenfunction that is positive in Ω, and is isolated in the spectrum $\sigma(-\Delta_p)$; see Anane [**9**] and Lindqvist [**68, 69**]. So the second eigenvalue $\lambda_2 = \inf \sigma(-\Delta_p) \cap (\lambda_1, \infty)$ is also defined; see Anane and Tsouli [**8**]. In the ODE case $n = 1$, where Ω is an interval, the spectrum consists of a sequence of simple eigenvalues $\lambda_k \nearrow \infty$ and the eigenfunction associated with λ_k has exactly $k - 1$ interior zeroes; see Cuesta [**35**] or Drábek [**46**]. In the semilinear PDE case $n \geqslant 2$, $p = 2$ also $\sigma(-\Delta)$ consists of a sequence of eigenvalues $\lambda_k \nearrow \infty$, but in the quasilinear PDE case $n \geqslant 2$, $p \neq 2$ a complete description of the spectrum is not available.

Eigenvalues of (0.2) are the critical values of the C^1-functional

$$I(u) = \int_\Omega |\nabla u|^p, \quad u \in S = \big\{u \in W = W_0^{1,p}(\Omega) : \|u\|_{L^p(\Omega)} = 1\big\},$$

which satisfies (PS). Denote by $\mathcal{A}$ the class of closed symmetric subsets of S and by

$$\gamma^+(A) = \sup \big\{k \geqslant 1 : \exists \text{ an odd continuous map } S^{k-1} \to A\big\},$$

$$\gamma^-(A) = \inf \big\{k \geqslant 1 : \exists \text{ an odd continuous map } A \to S^{k-1}\big\}$$

the genus and the cogenus of $A \in \mathcal{A}$, respectively, where S^{k-1} is the unit sphere in $\mathbb{R}^k$. Then

$$\lambda_k^\pm = \inf_{\substack{A \in \mathcal{A} \\ \gamma^\pm(A) \geqslant k}} \sup_{u \in A} I(u), \quad k \geqslant 1$$

are two increasing and unbounded sequences of eigenvalues, but, in general, it is not known whether either sequence is a complete list. The sequence (λ_k^+) was introduced by Drábek and Robinson [**47**]; γ_k^- is also called the Krasnosel'skii genus [**60**].

Solutions of (0.2) are the critical points of the functional

$$I_\lambda(u) = \int_\Omega |\nabla u|^p - \lambda\, |u|^p, \quad u \in W_0^{1,p}(\Omega).$$

When $\lambda \notin \sigma(-\Delta_p)$, the origin is the only critical point of I_λ and hence the critical groups $C_q(I_\lambda, 0)$ are defined. Again we take the coefficient group to be $\mathbb{Z}_2$. The following theorem is our main result on them.

Theorem 0.33 ([**98**, Proposition 1.1]). *The spectrum of $-\Delta_p$ contains a sequence of eigenvalues $\lambda_k \nearrow \infty$ such that $\lambda_k^- \leqslant \lambda_k \leqslant \lambda_k^+$ and*

$$\lambda \in (\lambda_k, \lambda_{k+1}) \backslash \sigma(-\Delta_p) \implies C_k(I_\lambda, 0) \neq 0.$$

Various applications of this sequence of eigenvalues can be found in Perera [**99, 100**], Liu and Li [**75**], Perera and Szulkin [**105**], Cingolani and Degiovanni [**30**], Guo and Liu [**55**], Degiovanni and Lancelotti [**43, 44**], Tanaka [**129**], Fang and Liu [**50**], Medeiros and Perera [**82**], Motreanu and Perera [**86**], and Degiovanni, Lancelotti, and Perera [**42**].

The eigenvalues λ_k are defined using the Yang index, whose definition and some properties we now recall. Yang [**133**] considered compact Hausdorff spaces with fixed-point-free continuous involutions and used the Čech homology theory, but for our purposes here it suffices to work with closed symmetric subsets of Banach spaces that do not contain the origin and singular homology groups.

Following [**133**], we first construct a special homology theory defined on the category of all pairs of closed symmetric subsets of Banach spaces that do not contain the origin and all continuous odd maps of such pairs. Let (X, A), $A \subset X$ be such a pair and $C(X, A)$ its singular chain complex with $\mathbb{Z}_2$-coefficients, and denote by $T_\#$ the chain map of $C(X, A)$ induced by the antipodal map $T(u) = -u$. We say that a q-chain c is symmetric if $T_\#(c) = c$, which holds if and only if $c = c' + T_\#(c')$ for some q-chain c'. The symmetric q-chains form a subgroup $C_q(X, A; T)$ of $C_q(X, A)$, and the boundary operator ∂_q maps $C_q(X, A; T)$ into $C_{q-1}(X, A; T)$, so these subgroups form a subcomplex $C(X, A; T)$. We denote by

$$Z_q(X, A; T) = \big\{c \in C_q(X, A; T) : \partial_q c = 0\big\},$$

$$B_q(X, A; T) = \big\{\partial_{q+1} c : c \in C_{q+1}(X, A; T)\big\},$$

$$H_q(X, A; T) = Z_q(X, A; T)/B_q(X, A; T)$$

the corresponding cycles, boundaries, and homology groups. A continuous odd map $f : (X, A) \to (Y, B)$ of pairs as above induces a chain map $f_\# : C(X, A; T) \to C(Y, B; T)$ and hence homomorphisms

$$f_* : H_q(X, A; T) \to H_q(Y, B; T).$$

For example,

$$H_q(S^k; T) = \begin{cases} \mathbb{Z}_2, & 0 \leqslant q \leqslant k \\ 0, & q \geqslant k + 1 \end{cases}$$

(see [**133**, Example 1.8]).

Let X be as above, and define homomorphisms $\nu : Z_q(X; T) \to \mathbb{Z}_2$ inductively by

$$\nu(z) = \begin{cases} \operatorname{In}(c), & q = 0 \\ \nu(\partial c), & q > 0 \end{cases}$$

if $z = c + T_\#(c)$, where the index of a 0-chain $c = \sum_i n_i \sigma_i$ is defined by $\operatorname{In}(c) = \sum_i n_i$. As in [**133**], ν is well-defined and $\nu\, B_q(X; T) = 0$, so we can define the index homomorphism $\nu_* : H_q(X; T) \to \mathbb{Z}_2$ by $\nu_*([z]) = \nu(z)$. If F is a closed subset of X such that $F \cup T(F) = X$ and $A = F \cap T(F)$, then there is a homomorphism

$$\Delta : H_q(X; T) \to H_{q-1}(A; T)$$

such that $\nu_*(\Delta[z]) = \nu_*([z])$ (see [**133**, Proposition 2.8]). Taking $F = X$ we see that if $\nu_*\, H_k(X; T) = \mathbb{Z}_2$, then $\nu_*\, H_q(X; T) = \mathbb{Z}_2$ for $0 \leqslant q \leqslant k$. We define the Yang index of X by

$$i(X) = \inf \{k \geqslant -1 : \nu_* \, H_{k+1}(X;T) = 0\},$$

taking $\inf \varnothing = \infty$. Clearly, $\nu_* \, H_0(X;T) = \mathbb{Z}_2$ if $X \neq \varnothing$, so $i(X) = -1$ if and only if $X = \varnothing$. For example, $i(S^k) = k$ (see [**133**, Example 3.4]).

Proposition 0.34 ([**133**, Proposition 2.4]). *If $f : X \to Y$ is as above, then $\nu_*(f_*([z])) = \nu_*([z])$ for $[z] \in H_q(X;T)$, and hence $i(X) \leqslant i(Y)$. In particular, this inequality holds if $X \subset Y$.*

Thus, $k^+ - 1 \leqslant i(X) \leqslant k^- - 1$ if there are odd continuous maps $S^{k^+-1} \to X \to S^{k^--1}$, so

$$\gamma^+(X) \leqslant i(X) + 1 \leqslant \gamma^-(X). \tag{0.3}$$

Proposition 0.35 ([**98**, Proposition 2.6]). *If $i(X) = k \geqslant 0$, then $\widetilde{H}_k(X) \neq 0$.*

Proof. We have

$$\nu_* \, H_q(X;T) = \begin{cases} \mathbb{Z}_2, & 0 \leqslant q \leqslant k \\ 0, & q \geqslant k+1. \end{cases}$$

We show that if $[z] \in H_k(X;T)$ is such that $\nu_*([z]) \neq 0$, then $[z] \neq 0$ in $\widetilde{H}_k(X)$. Arguing indirectly, assume that $z \in B_k(X)$, say, $z = \partial c$. Since $z \in B_k(X;T)$, $T_\#(z) = z$. Let $c' = c + T_\#(c)$. Then $c' \in Z_{k+1}(X;T)$ since $\partial c' = z + T_\#(z) = 2z = 0 \mod 2$, and $\nu_*([c']) = \nu(c') = \nu(\partial c) = \nu(z) \neq 0$, contradicting $\nu_* \, H_{k+1}(X;T) = 0$. □

Lemma 0.36. *We have*

$$C_q(I_\lambda, 0) \approx \widetilde{H}_{q-1}(I^\lambda) \quad \forall q.$$

Proof. Taking $U = \{u \in W : \|u\|_{L^p(\Omega)} \leqslant 1\}$ in (0.1) gives

$$C_q(I_\lambda, 0) = H_q(I_\lambda^0 \cap U, I_\lambda^0 \cap U \setminus \{0\}).$$

Since I_λ is positive homogeneous, $I_\lambda^0 \cap U$ radially contracts to the origin via

$$(I_\lambda^0 \cap U) \times [0,1] \to I_\lambda^0 \cap U, \quad (u,t) \mapsto (1-t)\,u$$

and $I_\lambda^0 \cap U \setminus \{0\}$ deformation retracts onto $I_\lambda^0 \cap S$ via

$$(I_\lambda^0 \cap U \setminus \{0\}) \times [0,1] \to I_\lambda^0 \cap U \setminus \{0\}, \quad (u,t) \mapsto (1-t)\,u + t\,u/\|u\|_{L^p(\Omega)},$$

so it follows from the exact sequence of the pair $(I_\lambda^0 \cap U, I_\lambda^0 \cap U \setminus \{0\})$ that

$$H_q(I_\lambda^0 \cap U, I_\lambda^0 \cap U \setminus \{0\}) \approx \widetilde{H}_{q-1}(I_\lambda^0 \cap S).$$

Since $I_\lambda|_S = I - \lambda$, $I_\lambda^0 \cap S = I^\lambda$. □

We are now ready to prove Theorem 0.33.

PROOF OF THEOREM 0.33. Set

$$\lambda_k = \inf_{\substack{A \in \mathcal{A} \\ i(A) \geqslant k-1}} \sup_{u \in A} I(u), \quad k \geqslant 1.$$

Then (λ_k) is an increasing sequence of critical points of I, and hence eigenvalues of $-\Delta_p$, by a standard deformation argument (see [**98**, Proposition 3.1]). By (0.3), $\lambda_k^- \leqslant \lambda_k \leqslant \lambda_k^+$, in particular, $\lambda_k \to \infty$.

Let $\lambda \in (\lambda_k, \lambda_{k+1}) \backslash \sigma(-\Delta_p)$. By Lemma 0.36, $C_k(I_\lambda, 0) \approx \widetilde{H}_{k-1}(I^\lambda)$, and $I^\lambda \in \mathcal{A}$ since I is even. Since $\lambda > \lambda_k$, there is an $A \in \mathcal{A}$ with $i(A) \geqslant k-1$ such that $I \leqslant \lambda$ on A. Then $A \subset I^\lambda$ and hence $i(I^\lambda) \geqslant i(A) \geqslant k-1$ by Proposition 0.34. On the other hand, $i(I^\lambda) \leqslant k-1$ since $I \leqslant \lambda < \lambda_{k+1}$ on I^λ. So $i(I^\lambda) = k-1$ and hence $\widetilde{H}_{k-1}(I^\lambda) \neq 0$ by Proposition 0.35. $\square$

CHAPTER 1

Abstract Formulation and Examples

Recall that an operator $q : W \to W'$, from one Banach space to another, is s-homogeneous, where $s > 0$, if

$$q(\alpha u) = \alpha^s \, q(u) \quad \forall u \in W, \, \alpha \geqslant 0.$$

Taking $\alpha = 0$ gives $q(0) = 0$. Note that q is s-homogeneous and odd if and only if

$$q(\alpha u) = |\alpha|^{s-1} \, \alpha \, q(u) \quad \forall u \in W, \, \alpha \in \mathbb{R},$$

and s-homogeneous and even if and only if

$$q(\alpha u) = |\alpha|^s \, q(u) \quad \forall u \in W, \, \alpha \in \mathbb{R}.$$

A continuous homogeneous operator is bounded in the following sense.

Proposition 1.1. *If $q \in C(W, W')$ is s-homogeneous, then there is a constant $C > 0$ such that*

$$\|q(u)\| \leqslant C \, \|u\|^s \quad \forall u \in W.$$

In particular, q is bounded on bounded subsets of W.

PROOF. If not, there is a sequence $(u_j) \subset W \backslash \{0\}$ such that

$$\|q(u_j)\| > j^{s+1} \, \|u_j\|^s \quad \forall j.$$

Let $\widehat{u}_j = u_j / (j \, \|u_j\|)$. Then $\|\widehat{u}_j\| = 1/j \to 0$, but

$$\|q(\widehat{u}_j)\| = \frac{\|q(u_j)\|}{j^s \, \|u_j\|^s} > j \to \infty,$$

contradicting the continuity of q. □

Let $L(W, W')$ be the Banach space of bounded linear operators from W to W'. Then $q : U \to W'$, where U is an open set in W, is differentiable at $u \in U$ if there is a $q'(u) \in L(W, W')$, called the derivative of q at u, such that

$$q(u + v) = q(u) + q'(u) \, v + \mathrm{o}(\|v\|) \quad \text{as} \quad v \to 0 \quad \text{in} \quad W.$$

The map q is continuously differentiable on U, or belongs to the class $C^1(U, W')$, if $q' \in C(U, L(W, W'))$, i.e., q' is defined everywhere in U and $U \to L(W, W')$, $u \mapsto q'(u)$ is continuous.

Let W^* be the dual of W and $\langle \cdot, \cdot \rangle$ the pairing between W^* and W. Then $q \in C(W, W^*)$ is a potential operator if there is a functional $Q \in C^1(W, \mathbb{R})$, called a potential for q, such that

$$Q'(u) = q(u) \quad \forall u \in W.$$

Replacing Q with $Q - Q(0)$ if necessary, we may assume that $Q(0) = 0$.

Proposition 1.2. *If q is a potential operator and Q is its potential satisfying $Q(0) = 0$, then*

$$Q(u) = \int_0^1 \langle q(tu), u \rangle \, dt \quad \forall u \in W.$$

In particular, Q is even if q is odd. If q is s-homogeneous, then

$$Q(u) = \frac{1}{s+1} \langle q(u), u \rangle \quad \forall u \in W$$

and is $(s+1)$-homogeneous.

PROOF. We have

$$Q(u) = \int_0^1 \frac{d}{dt} \left(Q(tu) \right) dt = \int_0^1 \langle Q'(tu), u \rangle \, dt = \int_0^1 \langle q(tu), u \rangle \, dt.$$

The last integral equals

$$\int_0^1 t^s \langle q(u), u \rangle \, dt = \frac{1}{s+1} \langle q(u), u \rangle$$

when q is s-homogeneous. □

We consider the nonlinear operator equation

$$A_p u = f(u) \tag{1.1}$$

in the dual $(W^*, \|\cdot\|^*)$ of a real reflexive Banach space $(W, \|\cdot\|)$ of dimension d, $1 \leqslant d \leqslant \infty$, where $A_p \in C(W, W^*)$ is

(A_1) $(p-1)$-homogeneous and odd for some $p \in (1, \infty)$,

(A_2) uniformly positive: $\exists\, c_0 > 0$ such that

$$\langle A_p u, u \rangle \geqslant c_0 \|u\|^p \quad \forall u \in W,$$

(A_3) a potential operator,

(A_4) of type (S): every sequence $(u_j) \subset W$ such that

$$u_j \rightharpoonup u, \quad \langle A_p u_j, u_j - u \rangle \to 0$$

has a subsequence that converges strongly to u,

and $f : W \to W^*$ is a compact potential operator.

By Proposition 1.2, the potential I_p of A_p satisfying $I_p(0) = 0$ is given by

$$I_p(u) = \frac{1}{p} \langle A_p u, u \rangle \tag{1.2}$$

and is p-homogeneous and even, and the potential F of f satisfying $F(0) = 0$ is given by

$$F(u) = \int_0^1 \langle f(tu), u \rangle \, dt. \tag{1.3}$$

Let

$$\Phi(u) = I_p(u) - F(u) = \frac{1}{p} \langle A_p u, u \rangle - \int_0^1 \langle f(tu), u \rangle \, dt, \quad u \in W. \tag{1.4}$$

Then

$$\Phi'(u) = I_p'(u) - F'(u) = A_p u - f(u),$$

so the solutions of equation (1.1) coincide with the critical points of the C^1-functional Φ, i.e., $u \in W$ solves (1.1) if and only if $\Phi'(u) = 0$. We note that $\Phi(0) = 0$, and Φ is even when f is odd.

By Proposition 1.1, there is a constant $C_0 > 0$ such that

$$\|A_p u\|^* \leqslant C_0 \|u\|^{p-1} \quad \forall u \in W. \tag{1.5}$$

Combining (1.2), (A_2), and (1.5) gives

$$\frac{c_0}{p} \|u\|^p \leqslant I_p(u) \leqslant \frac{C_0}{p} \|u\|^p \quad \forall u \in W. \tag{1.6}$$

In particular, $I_p \geqslant 0$ and is bounded on bounded sets. Since the compact operator f maps bounded sets into precompact, and hence bounded, sets, F is also bounded on bounded sets by (1.3). Therefore Φ is bounded on bounded subsets of W.

In many important applications

$$\langle A_p u, v \rangle \leqslant r \|u\|^{p-1} \|v\|, \quad \langle A_p u, u \rangle = r \|u\|^p \quad \forall u, v \in W \tag{1.7}$$

for some $r > 0$, or this can be achieved with respect to an equivalent norm.

Proposition 1.3. *If W is uniformly convex and (1.7) holds, then (A_4) is satisfied.*

Proof. If $u_j \rightharpoonup u$ and $\langle A_p u_j, u_j - u \rangle \to 0$, then

$$\begin{aligned} 0 &\leqslant r \left(\|u_j\|^{p-1} - \|u\|^{p-1} \right) \left(\|u_j\| - \|u\| \right) \\ &\leqslant \langle A_p u_j, u_j \rangle - \langle A_p u_j, u \rangle - \langle A_p u, u_j \rangle + \langle A_p u, u \rangle \\ &= \langle A_p u_j, u_j - u \rangle - \langle A_p u, u_j - u \rangle \to 0, \end{aligned} \tag{1.8}$$

so $\|u_j\| \to \|u\|$ and hence $u_j \to u$ by uniform convexity. □

Next we give some examples. In all the following examples (A_1) - (A_3) are clear and (A_4) follows from Proposition 1.3 and Hölder inequalities.

1.1. p-Laplacian Problems

Our motivating example is the p-Laplacian problem with the Dirichlet boundary condition

$$\begin{cases} -\Delta_p\, u = f(x,u) & \text{in } \Omega \\ \\ u = 0 & \text{on } \partial\Omega \end{cases} \tag{1.9}$$

where Ω is a bounded domain in $\mathbb{R}^n$, $n \geqslant 1$, $\Delta_p\, u = \operatorname{div}\left(|\nabla u|^{p-2}\, \nabla u\right)$ is the p-Laplacian of u, $p \in (1, \infty)$, and the nonlinearity f is in the space $\operatorname{Car}(\Omega \times \mathbb{R})$ of Carathéodory functions on $\Omega \times \mathbb{R}$ and has subcritical growth, i.e.,

$$|f(x,t)| \leqslant C\left(|t|^{r-1} + 1\right) \quad \forall (x,t) \in \Omega \times \mathbb{R} \tag{1.10}$$

for some $r \in (1, p^*)$, where

$$p^* = \begin{cases} \dfrac{np}{n-p}, & p < n \\ \\ \infty, & p \geqslant n \end{cases}$$

is the critical Sobolev exponent, and $C > 0$.

Here W is the Sobolev space $W_0^{1,\,p}(\Omega)$ with the usual norm

$$\|u\| = \left(\int_\Omega |\nabla u|^p\right)^{\frac{1}{p}}$$

and the dual $W^{-1,\,p'}(\Omega)$ where $p' = p/(p-1)$ is the Hölder conjugate of p,

$$\langle A_p\, u, v\rangle = \int_\Omega |\nabla u|^{p-2}\, \nabla u \cdot \nabla v, \qquad I_p(u) = \frac{1}{p}\int_\Omega |\nabla u|^p,$$

and

$$\langle f(u), v\rangle = \int_\Omega f(x,u)\, v, \qquad F(u) = \int_\Omega F(x,u) \tag{1.11}$$

where

$$F(x,t) = \int_0^t f(x,s)\, ds. \tag{1.12}$$

The operator f is compact by the growth condition (1.10) and the compactness of the Sobolev imbedding $W_0^{1,\,p}(\Omega) \hookrightarrow L^r(\Omega)$ for $r < p^*$.

We will use this example throughout the text to illustrate our results for equation (1.1). In the following sections we give some generalizations.

1.2. A_p-Laplacian Problems

Problem (1.9) is the special case $a_{ij}(x) \equiv \delta_{ij}$ of the A_p-Laplacian problem

$$\begin{cases} -A_p\, u = f(x,u) & \text{in } \Omega \\ \\ u = 0 & \text{on } \partial\Omega \end{cases}$$

where

$$\begin{aligned} A_p\, u &= \sum_{i,j=1}^{n} \frac{\partial}{\partial x_i} \left[\left(\sum_{k,l=1}^{n} a_{kl}(x)\, \frac{\partial u}{\partial x_k}\, \frac{\partial u}{\partial x_l} \right)^{\frac{p-2}{2}} a_{ij}(x)\, \frac{\partial u}{\partial x_j} \right] \\ &= \operatorname{div}\left(|\nabla u|_{A(x)}^{p-2}\, A(x) \nabla u \right) \end{aligned}$$

is the A_p-Laplacian of u, each $a_{ij} = a_{ji} \in L^\infty(\Omega)$, and the matrix $A = (a_{ij})$ satisfies

$$|\xi|_{A(x)}^2 := A(x)\, \xi \cdot \xi \geqslant |\xi|^2 \quad \forall \xi \in \mathbb{R}^n, \text{ a.e.}$$

Now we use the norm

$$\|u\| = \left(\int_\Omega |\nabla u|_{A(x)}^p \right)^{\frac{1}{p}}$$

in $W_0^{1,p}(\Omega)$ and take

$$\langle A_p\, u, v\rangle = \int_\Omega |\nabla u|_{A(x)}^{p-2}\, A(x) \nabla u \cdot \nabla v, \qquad I_p(u) = \frac{1}{p} \int_\Omega |\nabla u|_{A(x)}^p.$$

We note that the special case $a_{ij}(x) \equiv a(x)^{\frac{2}{p}}\, \delta_{ij}$, $1 \leqslant a \in L^\infty(\Omega)$ is the weighted p-Laplacian problem

$$\begin{cases} -\operatorname{div}\left(a(x)\, |\nabla u|^{p-2}\, \nabla u\right) = f(x,u) & \text{in } \Omega \\ u = 0 & \text{on } \partial\Omega. \end{cases} \tag{1.13}$$

1.3. Problems in Weighted Sobolev Spaces

Now we weaken the assumptions on a in problem (1.13) to $a(x) > 0$ a.e., $a \in L^1_{\mathrm{loc}}(\Omega)$, and $a^{-s} \in L^1(\Omega)$ for some $s \geqslant \max\{1/(p-1), n/p\}$, for example,

$$a(x) = 1/\operatorname{dist}(x, \partial\Omega).$$

Let $W = W_{0,a}^{1,p}(\Omega)$ be the completion of $C_0^\infty(\Omega)$ under the norm

$$\|u\| = \left(\int_\Omega a(x)\, |\nabla u|^p \right)^{\frac{1}{p}}.$$

By the Hölder inequality,

$$\int_\Omega |\nabla u|^{p_s} \leqslant \left(\int_\Omega a(x)^{-s} \right)^{\frac{1}{s+1}} \left(\int_\Omega a(x)\, |\nabla u|^p \right)^{\frac{s}{s+1}}, \quad u \in W$$

where $p_s = ps/(s+1) \geqslant 1$ since $s \geqslant 1/(p-1)$. So

$$W \hookrightarrow W_0^{1,p_s}(\Omega) \hookrightarrow L^r(\Omega) \tag{1.14}$$

for

$$r \begin{cases} \leqslant p_s^* = \dfrac{nps}{ns - (ps - n)}, & ps < n \\ < p_s^* = \infty, & ps \geqslant n, \end{cases}$$

in particular, $W \hookrightarrow L^p(\Omega)$ since $s \geqslant n/p$, and the last imbedding in (1.14) is compact for $r < p_s^*$. Thus, we assume that f satisfies (1.10) with $r \in (1, p_s^*)$ and take

$$\langle A_p\, u, v\rangle = \int_\Omega a(x)\, |\nabla u|^{p-2}\, \nabla u \cdot \nabla v, \qquad I_p(u) = \frac{1}{p} \int_\Omega a(x)\, |\nabla u|^p.$$

1.4. q-Kirchhoff Problems

Problem (1.9) is also the special case $q = p$ of the q-Kirchhoff type nonlocal problem

$$\begin{cases} -\left(\displaystyle\int_\Omega |\nabla u|^q\right)^{\frac{p-q}{q}} \Delta_q\, u = f(x, u) & \text{in } \Omega \\ \qquad\qquad u = 0 & \text{on } \partial\Omega \end{cases}$$

where $q \in (1, p]$ and f satisfies (1.10) with $r \in (1, q^*)$. Now $W = W_0^{1,q}(\Omega)$ and

$$\langle A_p\, u, v\rangle = \left(\int_\Omega |\nabla u|^q\right)^{\frac{p-q}{q}} \int_\Omega |\nabla u|^{q-2}\, \nabla u \cdot \nabla v,$$

$$I_p(u) = \frac{1}{p} \left(\int_\Omega |\nabla u|^q\right)^{\frac{p}{q}}.$$

1.5. Dynamic Equations on Time Scales

When $n = 1$ in problem (1.9), we may also consider the corresponding dynamic equation

$$\text{(1.15)} \qquad \begin{cases} -\left(|u^\Delta(t)|^{p-2}\, u^\Delta(t)\right)^\Delta = f(t, u^\sigma(t)), & t \in (a, b) \cap \mathbb{T} \\ \qquad u(a) = u(b) = 0 \end{cases}$$

where $\mathbb{T}$, called a time scale, is a nonempty closed subset of $[a, b]$,

$$\sigma(t) = \inf \left\{s \in \mathbb{T} : s > t\right\}$$

is the forward jump operator,

$$u^\Delta(t) = \lim_{\substack{s \to t \\ s \neq \sigma(t)}} \frac{u(\sigma(t)) - u(s)}{\sigma(t) - s}$$

is the Δ-derivative of u, and

$$u^\sigma(t) = u(\sigma(t)).$$

In particular, (1.15) is an ordinary differential equation when $\mathbb{T}$ is continuous and a difference equation when $\mathbb{T}$ is discrete. Now W is the space $W^{1,p}_{0,\Delta}(a,b)$, with the norm

$$\|u\| = \left(\int_a^b |u^\Delta(t)|^p \, \Delta t\right)^{\frac{1}{p}},$$

constructed by Agarwal, Otero-Espinar, Perera, and Vivero [**2**],

$$\langle A_p \, u, v\rangle = \int_a^b |u^\Delta(t)|^{p-2} \, u^\Delta(t) \, v^\Delta(t) \, \Delta t, \qquad I_p(u) = \frac{1}{p}\int_a^b |u^\Delta(t)|^p \, \Delta t,$$

and

$$\langle f(u), v\rangle = \int_a^b f(t, u^\sigma(t)) \, v^\sigma(t) \, \Delta t, \qquad F(u) = \int_a^b F(t, u^\sigma(t)) \, \Delta t$$

where $\int_a^b \cdot \, \Delta t$ is the Lebesgue Δ-integral and F is the primitive of f given by (1.12). We refer to Agarwal, Bohner, O'Regan, and Peterson [**1**] for a survey of dynamic equations on time scales.

1.6. Other Boundary Conditions

Our operator setting also includes problems with other boundary conditions.

Nonlinear Flux Boundary Conditions. Consider

$$\begin{cases} -\Delta_p \, u + a(x) \, |u|^{p-2} \, u = f(x,u) & \text{in } \Omega \\ |\nabla u|^{p-2} \, \dfrac{\partial u}{\partial \nu} = b(x,u) & \text{on } \partial\Omega \end{cases} \tag{1.16}$$

where $\partial\Omega$ is now assumed to be C^1, $a \in L^\infty(\Omega)$ with $\operatorname{ess\,inf} a > 0$, f is as in Section 1.1, $\partial/\partial\nu$ is the exterior normal derivative on $\partial\Omega$, and $b \in \operatorname{Car}(\partial\Omega \times \mathbb{R})$ satisfies the growth condition

$$|b(x,t)| \leqslant C\left(|t|^{s-1} + 1\right) \quad \forall (x,t) \in \partial\Omega \times \mathbb{R} \tag{1.17}$$

for some $s \in (1, p_1^*)$ where

$$p_1^* = \begin{cases} \dfrac{(n-1)\,p}{n-p}, & p < n \\ \infty, & p \geqslant n \end{cases}$$

is the critical Sobolev trace exponent. In particular, (1.16) is the Neumann problem when $b = 0$.

Here $W = W^{1,p}(\Omega)$ with the norm

$$\|u\| = \left(\int_\Omega |\nabla u|^p + a(x) \, |u|^p\right)^{\frac{1}{p}}, \tag{1.18}$$

$$\langle A_p\, u, v\rangle = \int_\Omega |\nabla u|^{p-2}\, \nabla u \cdot \nabla v + a(x)\, |u|^{p-2}\, uv, \tag{1.19}$$

$$I_p(u) = \frac{1}{p} \int_\Omega |\nabla u|^p + a(x)\, |u|^p,$$

and

$$\langle f(u), v\rangle = \int_\Omega f(x,u)\, v + \int_{\partial\Omega} b(x,u)\, v, \tag{1.20}$$

$$F(u) = \int_\Omega F(x,u) + \int_{\partial\Omega} B(x,u)$$

where

$$B(x,t) = \int_0^t b(x,s)\, ds.$$

The operator f is still compact by (1.17) and the compactness of the Sobolev trace imbedding $W^{1,p}(\Omega) \hookrightarrow L^s(\partial\Omega)$ for $s < p_1^*$.

Let us verify the first inequality in (1.7) since here it is not as straightforward as in the other examples. By the Hölder inequalities for integrals and sums,

$$\begin{aligned}
\langle A_p\, u, v\rangle &\leqslant \left(\int_\Omega |\nabla u|^p\right)^{\frac{1}{p'}} \left(\int_\Omega |\nabla v|^p\right)^{\frac{1}{p}} \\
&\quad + \left(\int_\Omega a(x)\, |u|^p\right)^{\frac{1}{p'}} \left(\int_\Omega a(x)\, |v|^p\right)^{\frac{1}{p}} \\
&\leqslant \left(\int_\Omega |\nabla u|^p + a(x)\, |u|^p\right)^{\frac{1}{p'}} \left(\int_\Omega |\nabla v|^p + a(x)\, |v|^p\right)^{\frac{1}{p}} \\
&= \|u\|^{p-1}\, \|v\|\, .
\end{aligned}$$

Periodic Boundary Conditions. Let $\Omega = [0, L_1] \times \cdots \times [0, L_n]$. We say that a function on Ω is periodic if it is L_i-periodic in x_i for $i = 1, \ldots, n$. Consider

$$\begin{cases} -\Delta_p\, u + a(x)\, |u|^{p-2}\, u = f(x,u) & \text{in } \Omega \\ \qquad\qquad u \text{ is periodic} \end{cases}$$

where $a \in L^\infty(\Omega)$ is periodic with $\operatorname{ess\,inf} a > 0$ and $f \in \operatorname{Car}(\Omega \times \mathbb{R})$ is periodic in x with subcritical growth. Now $W = W^{1,p}_{\text{per}}(\Omega) \subset W^{1,p}(\Omega)$ is the completion of the space $C^\infty_{\text{per}}(\Omega)$ of smooth periodic functions on Ω under the norm (1.18), A_p and I_p are still given by (1.19), and f and F by (1.20) with $b = 0$.

1.7. p-Biharmonic Problems

We may also consider higher-order problems, for example, the fourth order p-biharmonic problem

$$\begin{cases} \Delta_p^2 u = f(x,u) & \text{in } \Omega \\ u = \nabla u = 0 & \text{on } \partial\Omega \end{cases}$$

where $\partial\Omega$ is now C^2, $\Delta_p^2 u = \Delta\big(|\Delta u|^{p-2}\, \Delta u\big)$ is the p-biharmonic operator, and f satisfies (1.10) with $r \in (1, p_2^*)$ where

$$p_2^* = \begin{cases} \dfrac{np}{n-2p}, & p < n/2 \\ \infty, & p \geqslant n/2 \end{cases}$$

is the critical exponent for the Sobolev imbedding $W^{2,p}(\Omega) \hookrightarrow L^r(\Omega)$.

Here $W = W_0^{2,p}(\Omega)$ with the norm

$$\|u\| = \left(\int_\Omega |\Delta u|^p\right)^{\frac{1}{p}}$$

and

$$\langle A_p\, u, v\rangle = \int_\Omega |\Delta u|^{p-2}\, \Delta u\, \Delta v, \qquad I_p(u) = \frac{1}{p}\int_\Omega |\Delta u|^p.$$

The operator f defined in (1.11) is compact by the compactness of the imbedding $W_0^{2,p}(\Omega) \hookrightarrow L^r(\Omega)$ for $r < p_2^*$.

We may consider the p-biharmonic problem with the Navier boundary conditions

$$\begin{cases} \Delta_p^2 u = f(x,u) & \text{in } \Omega \\ u = \Delta u = 0 & \text{on } \partial\Omega \end{cases}$$

by working in the space $W = W^{2,p}(\Omega) \cap W_0^{1,p}(\Omega)$ with the same norm, etc.

1.8. Systems of Equations

We may also consider the systems corresponding to the problems considered in the preceding sections, for example, the p-Laplacian system

$$\begin{cases} -\Delta_p\, u = \nabla_u\, F(x,u) & \text{in } \Omega \\ u = 0 & \text{on } \partial\Omega \end{cases} \tag{1.21}$$

where $u = (u_1, \ldots, u_m)$, $\Delta_p\, u = (\Delta_p\, u_1, \ldots, \Delta_p\, u_m)$, and $F \in C^1(\Omega \times \mathbb{R}^m)$ satisfies $F(x,0) \equiv 0$ and

$$|\nabla_u\, F(x,u)| \leqslant C\left(|u|^{r-1} + 1\right) \quad \forall (x,u) \in \Omega \times \mathbb{R}^m,$$

where $\nabla_u F = (\partial F/\partial u_1, \dots, \partial F/\partial u_m)$, for some $r \in (1, p^*)$ and $C > 0$. Then $W = (W_0^{1,p}(\Omega))^m$, the product of m copies of $W_0^{1,p}(\Omega)$, with the norm

$$\|u\| = \left(\int_\Omega \sum_{i=1}^m |\nabla u_i|^p\right)^{\frac{1}{p}},$$

$$\langle A_p\, u, v\rangle = \int_\Omega \sum_{i=1}^m |\nabla u_i|^{p-2}\, \nabla u_i \cdot \nabla v_i, \qquad I_p(u) = \frac{1}{p}\int_\Omega \sum_{i=1}^m |\nabla u_i|^p,$$

and

$$\langle f(u), v\rangle = \int_\Omega \nabla_u\, F(x, u) \cdot v, \qquad F(u) = \int_\Omega F(x, u). \tag{1.22}$$

By Hölder inequalities,

$$\begin{aligned}\langle A_p\, u, v\rangle &\leqslant \sum_{i=1}^m \left(\int_\Omega |\nabla u_i|^p\right)^{\frac{1}{p'}} \left(\int_\Omega |\nabla v_i|^p\right)^{\frac{1}{p}} \\ &\leqslant \left(\int_\Omega \sum_{i=1}^m |\nabla u_i|^p\right)^{\frac{1}{p'}} \left(\int_\Omega \sum_{i=1}^m |\nabla v_i|^p\right)^{\frac{1}{p}} = \|u\|^{p-1}\, \|v\|,\end{aligned}$$

and clearly $\langle A_p\, u, u\rangle = \|u\|^p$, so (A_4) holds by Proposition 1.3.

CHAPTER 2

Background Material

In this preliminary chapter we collect some of the necessary background material on algebraic topology. Maps between topological spaces are always assumed to be continuous unless otherwise stated. We will use the following notations throughout the text.

For $1 \leqslant n \leqslant \infty$,

$$\mathbb{R}^n = \text{Euclidean } n\text{-space with the norm } \|x\| = \left(\sum_{i=1}^{n} x_i^2\right)^{1/2},$$

$$D^n = n\text{-disk} = \{x \in \mathbb{R}^n : \|x\| \leqslant 1\},$$

$$S^{n-1} = (n-1)\text{-sphere} = \partial D^n = \{x \in D^n : \|x\| = 1\},$$

$$\mathbb{R}\mathrm{P}^{n-1} = \text{real projective } (n-1)\text{-space} = \text{quotient space of } S^{n-1} \text{ with each } x \text{ and } -x \text{ identified.}$$

For $n, q \in \mathbb{Z}$,

$$\delta_{qn} = \text{Kronecker delta} = \begin{cases} 1, & q = n \\ 0, & q \neq n. \end{cases}$$

If A is a subset of a metric space (X, d),

$$\operatorname{dist}(x, A) = \text{distance from } x \text{ to } A = \inf_{x' \in A} d(x, x'),$$

$$N_\delta(A) = \delta\text{-neighborhood of } A = \{x \in X : \operatorname{dist}(x, A) \leqslant \delta\},$$

$$B_\delta(x) = (\text{closed}) \ \delta\text{-ball around } x = N_\delta(\{x\}) = \{x' \in X : d(x', x) \leqslant \delta\},$$

$$S_\delta(x) = \delta\text{-sphere around } x = \partial B_\delta(x) = \{x' \in X : d(x', x) = \delta\}.$$

If X is a topological space,

$$CX = \text{cone on } X = \text{quotient space of } X \times [0,1] \text{ with } X \times \{1\} \text{ collapsed to a point,}$$

$$SX = \text{suspension of } X = \text{quotient space of } X \times [-1,1] \text{ with } X \times \{1\} \text{ and } X \times \{-1\} \text{ collapsed to different points.}$$

When X is a subset of a normed linear space,

$$cX = \{cx : x \in X\}$$

for $c \in \mathbb{R}$ and

$$IX = \{tx : x \in X,\ t \in [0,1]\}.$$

If A is a subset of X,

$\overset{\circ}{A}$ = interior of A, $\quad \overline{A}$ = closure of A, $\quad A^c$ = complement of A in X.

2.1. Homotopy

We briefly recall the notions of homotopic maps and homotopy type of spaces.

A pair (X, A) of spaces consists of a topological space X and a subspace A of X. We set $(X, \varnothing) = X$. A map $f : (X, A) \to (Y, B)$ of pairs is a map $f : X \to Y$ such that $f(A) \subset B$.

Two maps of pairs $f_0, f_1 : (X, A) \to (Y, B)$ are homotopic, written $f_0 \simeq f_1$, if there is a map $F : X \times [0, 1] \to Y$, called a homotopy between f_0 and f_1, such that

(*i*) $F(\cdot, 0) = f_0$,
(*ii*) $F(\cdot, t) : (X, A) \to (Y, B)$ is a map of pairs for all $t \in [0, 1]$,
(*iii*) $F(\cdot, 1) = f_1$.

F is called a homotopy relative to A if $F(\cdot, t)|_A$ is independent of t. The homotopy class of $f : (X, A) \to (Y, B)$, denoted by $[f]$, is the set of maps homotopic to f. The set of all homotopy classes of maps from (X, A) to (Y, B) is denoted by $[(X, A), (Y, B)]$.

Two pairs of spaces (X, A) and (Y, B) are homotopy equivalent, or have the same homotopy type, written $(X, A) \simeq (Y, B)$, if there are maps $f : (X, A) \to (Y, B)$ and $g : (Y, B) \to (X, A)$, called homotopy equivalences between (X, A) and (Y, B), such that

$$fg \simeq id_{(Y,B)}, \qquad gf \simeq id_{(X,A)}.$$

A retraction of X onto A is a map $r : X \to A$ such that $r|_A = id_A$. A deformation retraction of X onto A is a homotopy relative to A between id_X and a retraction $r : X \to A$, i.e., a map $F : X \times [0, 1] \to X$ such that

(*i*) $F(\cdot, 0) = id_X$,
(*ii*) $F(\cdot, t)|_A = id_A \quad \forall t \in [0, 1]$,
(*iii*) $F(X, 1) = A$.

Then A is called a deformation retract of X, and since $ri = id_A$ and $ir \simeq id_X$ where $i : A \subset X$ is the inclusion, $X \simeq A$.

A space is contractible if it has the homotopy type of a one-point space, in particular, if it deformation retracts to a point.

Example 2.1. $\mathbb{R}^n$, D^n, and S^∞ are all contractible.

Triples (X, A, B) of spaces $X \supset A \supset B$ give rise to similar homotopy notions.

2.2. Direct Limits

We briefly recall the notions of directed systems and their limits.

A partial order on a set Λ is a relation $\preceq$ in Λ that is

(*i*) reflexive: $\lambda \preceq \lambda \quad \forall \lambda \in \Lambda$,
(*ii*) transitive: $\lambda \preceq \mu \preceq \nu \implies \lambda \preceq \nu$.

A partially ordered set Λ is directed if for each pair $\lambda, \mu \in \Lambda$, there is a $\nu \in \Lambda$ with $\lambda \preceq \nu$ and $\mu \preceq \nu$.

A directed system (of vector spaces) is a collection $\{V_\lambda\}_{\lambda \in \Lambda}$ of vector spaces indexed by a directed set Λ and linear maps $L_{\lambda\mu} : V_\lambda \to V_\mu$ for each pair $\lambda \preceq \mu$ such that

(*i*) $L_{\lambda\lambda} = id_{V_\lambda} \quad \forall \lambda \in \Lambda$,
(*ii*) $\lambda \preceq \mu \preceq \nu \implies L_{\lambda\nu} = L_{\mu\nu} L_{\lambda\mu}$.

Define an equivalence relation $\sim$ in the disjoint union of all the V_λ by setting $v_\lambda \in V_\lambda$ equivalent to $v_\mu \in V_\mu$ if there is a $\nu \in \Lambda$ with $L_{\lambda\nu} v_\lambda = L_{\mu\nu} v_\mu$. In particular, if $\lambda \preceq \mu$, then $v_\lambda \sim L_{\lambda\mu} v_\lambda$ since $L_{\lambda\mu} v_\lambda = L_{\mu\mu}(L_{\lambda\mu} v_\lambda)$. Clearly, $v_\lambda \sim v_\lambda$, and $v_\lambda \sim v_\mu$ implies $v_\mu \sim v_\lambda$. If $L_{\lambda\lambda'} v_\lambda = L_{\mu\lambda'} v_\mu$ and $L_{\mu\mu'} v_\mu = L_{\nu\mu'} v_\nu$, taking $\nu' \in \Lambda$ with $\lambda' \preceq \nu'$ and $\mu' \preceq \nu'$, we have

$$\begin{aligned} L_{\lambda\nu'} v_\lambda = L_{\lambda'\nu'} L_{\lambda\lambda'} v_\lambda = L_{\lambda'\nu'} L_{\mu\lambda'} v_\mu = L_{\mu\nu'} v_\mu &= L_{\mu'\nu'} L_{\mu\mu'} v_\mu \\ &= L_{\mu'\nu'} L_{\nu\mu'} v_\nu = L_{\nu\nu'} v_\nu, \end{aligned}$$

so $v_\lambda \sim v_\mu$ and $v_\mu \sim v_\nu$ imply $v_\lambda \sim v_\nu$. The set V of equivalence classes, called the direct limit of the directed system $\{V_\lambda\}$ and denoted by

$$\varinjlim_{\Lambda} V_\lambda,$$

is a vector space with addition and scalar multiplication defined by

$$[v_\lambda] + [v_\mu] = [L_{\lambda\nu} v_\lambda + L_{\mu\nu} v_\mu], \qquad c\,[v_\lambda] = [c\,v_\lambda]$$

where $\nu \in \Lambda$ with $\lambda \preceq \nu$ and $\mu \preceq \nu$. The quotient map $L_\lambda : V_\lambda \to V$, $v_\lambda \mapsto [v_\lambda]$ is then linear for each $\lambda \in \Lambda$.

Proposition 2.2. *If* $L_\lambda v_\lambda = 0$, *then there is a* $\mu \in \Lambda$ *such that* $L_{\lambda\mu} v_\lambda = 0$.

PROOF. Since $[v_\lambda] = 0 \in V$ contains $0 \in V_\lambda$, $v_\lambda \sim 0$, so there is a $\mu \in \Lambda$ with $L_{\lambda\mu} v_\lambda = L_{\lambda\mu} 0 = 0 \in V_\mu$ by linearity. □

A subset $\Lambda' \subset \Lambda$ is cofinal in Λ if for each $\lambda \in \Lambda$, there is a $\lambda' \in \Lambda'$ with $\lambda \preceq \lambda'$. For each pair $\lambda', \mu' \in \Lambda'$, first taking $\nu \in \Lambda$ with $\lambda' \preceq \nu$ and $\mu' \preceq \nu$, and then taking $\nu' \in \Lambda'$ with $\nu \preceq \nu'$, we have $\lambda' \preceq \nu'$ and $\mu' \preceq \nu'$ by transitivity, so Λ' is also directed. For each $v_\lambda \in V_\lambda$, $[v_\lambda] = [v_{\lambda'}]$ where $\lambda' \in \Lambda'$ with $\lambda \preceq \lambda'$ and $v_{\lambda'} = L_{\lambda\lambda'} v_\lambda \in V_{\lambda'}$, so

$$\varinjlim_{\Lambda} V_\lambda = \varinjlim_{\Lambda'} V_{\lambda'}.$$

In particular, if Λ has a maximal element ω, $\{\omega\}$ is cofinal and hence $V = V_\omega$.

Finally suppose W is a vector space and $M_\lambda : V_\lambda \to W$, $\lambda \in \Lambda$ are linear maps such that for each pair $\lambda \preceq \mu$, $M_\mu L_{\lambda\mu} = M_\lambda$, i.e., the diagram

$$\begin{array}{ccc} V_\lambda & \xrightarrow{L_{\lambda\mu}} & V_\mu \\ & \searrow^{M_\lambda} & \downarrow M_\mu \\ & & W \end{array}$$

commutes. Define a map $M : V \to W$ by $M[v_\lambda] = M_\lambda v_\lambda$, so that for each $\lambda \in \Lambda$, $ML_\lambda = M_\lambda$, i.e., the diagram

$$\begin{array}{ccc} V_\lambda & \xrightarrow{L_{\lambda}} & V \\ & \searrow^{M_\lambda} & \downarrow M \\ & & W \end{array}$$

commutes. If $v_\lambda \sim v_\mu$, taking $\nu \in \Lambda$ with $L_{\lambda\nu} v_\lambda = L_{\mu\nu} v_\mu$, we have

$$M_\lambda v_\lambda = M_\nu L_{\lambda\nu} v_\lambda = M_\nu L_{\mu\nu} v_\mu = M_\mu v_\mu,$$

so M is well-defined. It is also linear:

$$M([v_\lambda] + [v_\mu]) = M_\nu (L_{\lambda\nu} v_\lambda + L_{\mu\nu} v_\mu) = M_\lambda v_\lambda + M_\mu v_\mu = M[v_\lambda] + M[v_\mu]$$

where $\nu \in \Lambda$ with $\lambda \preceq \nu$ and $\mu \preceq \nu$, and

$$M(c\,[v_\lambda]) = M_\lambda (c\, v_\lambda) = c\, M_\lambda v_\lambda = c\, M[v_\lambda].$$

We write

$$M = \varinjlim_{\Lambda} M_\lambda.$$

2.3. Alexander-Spanier Cohomology Theory

We recall the construction and the axioms of Alexander-Spanier cohomology with coefficients in the field $\mathbb{Z}_2 = \{0, 1\}$ of characteristic 2. A basic reference is Spanier [**126**].

We begin with the cohomology of cochain complexes. A cochain complex C over $\mathbb{Z}_2$ is a sequence

$$\cdots \longrightarrow C^{q-1} \xrightarrow{\delta^{q-1}} C^q \xrightarrow{\delta^q} C^{q+1} \longrightarrow \cdots$$

of vector spaces C^q over $\mathbb{Z}_2$ and linear maps δ^q, called coboundary operators, such that

$$\delta^q \delta^{q-1} = 0 \quad \forall q. \tag{2.1}$$

If $C'^q \subset C^q$ is a sequence of subspaces such that $\delta^q(C'^q) \subset C'^{q+1}$, then $C' = \left\{C'^q, \delta^q\right\}$ is itself a cochain complex, called a subcomplex of C. Passing to quotients, δ^q induces a linear map

$$\overline{\delta}^q : C^q/C'^q \to C^{q+1}/C'^{q+1}$$

such that $\overline{\delta}^q \overline{\delta}^{q-1} = 0$, so $C/C' = \left\{C^q/C'^q, \overline{\delta}^q\right\}$ is also a cochain complex, called the quotient complex of C by C'.

The elements of C^q are called q-cochains and those of the subspaces

$$Z^q(C) = \ker \delta^q, \qquad B^q(C) = \operatorname{im} \delta^{q-1}$$

are called q-cocycles and q-coboundaries, respectively. By (2.1), $B^q(C) \subset Z^q(C)$, and the quotient space

$$H^q(C) = Z^q(C)/B^q(C)$$

is called the q-th cohomology group of C. If $B^q(C) = Z^q(C)$, and hence $H^q(C) = 0$, for all q, then C is called an exact sequence, and cohomology is a measure of nonexactness.

A cochain map $f : C \to C'$ of complexes consists of a sequence

$$\begin{array}{ccccccccc} \cdots & \longrightarrow & C^{q-1} & \xrightarrow{\delta^{q-1}} & C^q & \xrightarrow{\delta^q} & C^{q+1} & \longrightarrow & \cdots \\ & & \downarrow{\scriptstyle f^{q-1}} & & \downarrow{\scriptstyle f^q} & & \downarrow{\scriptstyle f^{q+1}} & & \\ \cdots & \longrightarrow & C'^{q-1} & \xrightarrow{\delta'^{q-1}} & C'^q & \xrightarrow{\delta'^q} & C'^{q+1} & \longrightarrow & \cdots \end{array}$$

of linear maps such that

$$f^{q+1}\delta^q = \delta'^q f^q \quad \forall q.$$

Then $f^q(Z^q(C)) \subset Z^q(C')$ and $f^q(B^q(C)) \subset B^q(C')$, so f induces linear maps

$$f^* : H^*(C) \to H^*(C')$$

between the cohomology groups of C and C'. Moreover, $(gf)^* = f^*g^*$ for a composed mapping $C \xrightarrow{f} C' \xrightarrow{g} C''$ and $id_C^* = id_{H^*(C)}$.

Now let (X, A) be a pair of spaces. We will construct a cochain complex whose cohomology will be defined to be the Alexander-Spanier cohomology of (X, A).

For $q \geqslant 0$, let $C^q(X)$ be the vector space (over $\mathbb{Z}_2$) of all functions $\varphi : X^{q+1} \to \mathbb{Z}_2$, where X^{q+1} is the product of $q + 1$ copies of X, with addition and scalar multiplication defined pointwise, i.e.,

$$(\varphi + \varphi')(x) = \varphi(x) + \varphi'(x), \qquad (c\,\varphi)(x) = c\,(\varphi(x))$$

for $\varphi, \varphi' \in C^q(X)$, $c \in \mathbb{Z}_2$, and $x = (x_0, \ldots, x_q) \in X^{q+1}$. Define a coboundary operator $\delta^q : C^q(X) \to C^{q+1}(X)$ by

$$(\delta^q \varphi)(x_0, \ldots, x_{q+1}) = \sum_{i=0}^{q+1} \varphi(x_0, \ldots, \widehat{x}_i, \ldots, x_{q+1})$$

where the "hat" indicates that x_i is omitted. Then $\delta^q\delta^{q-1} = 0$, so $C(X) = \{C^q(X), \delta^q\}$ is a cochain complex.

We say that $\varphi \in C^q(X)$ is locally zero if there is an open covering $\mathcal{U}$ of X such that φ vanishes on any $(q + 1)$-tuple of X that lies in some $U \in \mathcal{U}$, i.e., φ vanishes on $\mathcal{U}^{q+1} := \bigcup_{U \in \mathcal{U}} U^{q+1}$. Let $C_0^q(X)$ be the subspace of $C^q(X)$ consisting of locally zero functions. If φ vanishes on $\mathcal{U}^{q+1}$, then $\delta^q \varphi$ vanishes on $\mathcal{U}^{q+2}$, so $C_0(X) = \{C_0^q(X), \delta^q\}$ is a subcomplex of $C(X)$. Let

$\overline{C}(X) = C(X)/C_0(X) = \{C^q(X)/C_0^q(X), \overline{\delta}^q\}$. A map $f : X \to Y$ of spaces induces a cochain map $f^\# : C(Y) \to C(X)$ by

$$(f^{\#q}\psi)(x_0, \ldots, x_q) = \psi(f(x_0), \ldots, f(x_q)), \quad \psi \in C^q(Y).$$

If $\psi \in C_0^q(Y)$ and $\mathcal{V}$ is an open covering of Y such that ψ vanishes on $\mathcal{V}^{q+1}$, then $\mathcal{U} = f^{-1}\mathcal{V}$ is an open covering of X and $f^{\#q}\psi$ vanishes on $\mathcal{U}^{q+1}$. So $f^\#$ maps $C_0(Y)$ into $C_0(X)$ and hence induces a cochain map $\overline{f}^\# : \overline{C}(Y) \to \overline{C}(X)$.

If $i : A \subset X$, then $\overline{C}(X, A) = \ker \overline{i}^\# = \{\ker \overline{i}^{\#q}, \overline{\delta}^q\}$ is a subcomplex of $\overline{C}(X)$. Let $C(X, A)$ be the subcomplex of $C(X)$ consisting of functions that are locally zero on A. Then we see that $C_0(X) \subset C(X, A)$ and $\overline{C}(X, A) = C(X, A)/C_0(X)$. Define the Alexander-Spanier cohomology of (X, A) by

$$H^*(X, A) = H^*(\overline{C}(X, A)).$$

We write $H^*(X, \varnothing) = H^*(X)$. If $f : (X, A) \to (Y, B)$ is a map of pairs, then $\overline{f}^\# : \overline{C}(Y, B) \to \overline{C}(X, A)$ is a cochain map and hence induces linear maps

$$f^* : H^*(Y, B) \to H^*(X, A)$$

such that $(gf)^* = f^*g^*$ for $(X, A) \xrightarrow{f} (Y, B) \xrightarrow{g} (Z, C)$ and $id^*_{(X,A)} = id_{H^*(X,A)}$. We refer to Spanier [**126**] for the proofs of the axioms for this cohomology theory.

(c_1) Homotopy axiom: If $f_0 \simeq f_1 : (X, A) \to (Y, B)$, then

$$f_0^* = f_1^* : H^*(Y, B) \to H^*(X, A).$$

(c_2) Exactness axiom: Each triple (X, A, B) has an exact sequence

$$\cdots \longrightarrow H^q(X, A) \xrightarrow{j^*} H^q(X, B) \xrightarrow{i^*} H^q(A, B) \xrightarrow{\delta} H^{q+1}(X, A) \longrightarrow \cdots$$

where $i : (A, B) \subset (X, B)$, $j : (X, B) \subset (X, A)$, and δ is called the connecting map.

(c_3) Excision axiom: If U is an open subset of X such that $\overline{U} \subset \overset{\circ}{A}$, then $j : (X \backslash U, A \backslash U) \subset (X, A)$ induces an isomorphism

$$j^* : H^*(X, A) \approx H^*(X \backslash U, A \backslash U).$$

(c_4) Dimension axiom: If X is a one-point space, then

$$H^q(X) \approx \delta_{q0}\,\mathbb{Z}_2. \tag{2.2}$$

Cohomology is a homotopy invariant, i.e.,

$$(X, A) \simeq (Y, B) \implies H^*(X, A) \approx H^*(Y, B).$$

To see this, let $f : (X, A) \to (Y, B)$ and $g : (Y, B) \to (X, A)$ be homotopy equivalences. By (c_1),

$$f^*g^* = (gf)^* = id^*_{(X,A)} = id_{H^*(X,A)}$$

and similarly $g^* f^* = id_{H^*(Y,B)}$, so $g^* = (f^*)^{-1}$. In particular, the cohomology of a contractible space is given by (2.2). If $r : X \to A$ is only a retraction and $i : A \subset X$, then $i^* r^* = (ri)^* = id_A^* = id_{H^*(A)}$ and hence r^* is one-to-one and i^* is onto.

Example 2.3. $H^q(S^{n-1}) \approx \delta_{q0}\, \mathbb{Z}_2 \oplus \delta_{q,n-1}\, \mathbb{Z}_2$

When $X \neq \varnothing$, it is sometimes more convenient to work with the reduced groups

$$\widetilde{H}^q(X) = \begin{cases} H^0(X)/\mathbb{Z}_2, & q = 0 \\ \\ H^q(X), & q \geqslant 1, \end{cases}$$

which are trivial in all dimensions for contractible spaces. They also fit into an exact sequence

$$\cdots \longrightarrow \widetilde{H}^{q-1}(X) \longrightarrow \widetilde{H}^{q-1}(A) \longrightarrow H^q(X, A) \longrightarrow \widetilde{H}^q(X) \longrightarrow \cdots$$

for a pair (X, A) with $A \neq \varnothing$. In particular, we have

Proposition 2.4. *Let (X, A) be a pair with X contractible.*

(i) *If $A = \varnothing$, then*

$$H^q(X, A) \approx \delta_{q0}\, \mathbb{Z}_2.$$

(ii) *If $A \neq \varnothing$, then*

$$H^q(X, A) \approx \widetilde{H}^{q-1}(A) \quad \forall q.$$

In particular, $H^0(X, A) = 0$.

Example 2.5. $\widetilde{H}^q(S^{n-1}) \approx \delta_{q,n-1}\, \mathbb{Z}_2$, $H^q(D^n, S^{n-1}) \approx \widetilde{H}^{q-1}(S^{n-1}) \approx \delta_{qn}\, \mathbb{Z}_2$

There is a product $\smile : H^p(X) \times H^q(X) \to H^{p+q}(X)$, called the cup product, giving $H^*(X)$ the structure of a graded ring. To define this product, first define a product $\smile : C^p(X) \times C^q(X) \to C^{p+q}(X)$ by

$$(\varphi \smile \psi)(x_0, \ldots, x_{p+q}) = \varphi(x_0, \ldots, x_p)\, \psi(x_p, \ldots, x_{p+q}).$$

If φ or ψ is locally zero, then so is $\varphi \smile \psi$, and hence $\smile$ induces a product $\smile : \overline{C}^p(X) \times \overline{C}^q(X) \to \overline{C}^{p+q}(X)$. It is easy to verify that

$$\delta\,(\varphi \smile \psi) = \delta\,\varphi \smile \psi + \varphi \smile \delta\,\psi,$$

so if φ and ψ are cocycles, then so is $\varphi \smile \psi$, and if one of them is a coboundary in addition, then so is $\varphi \smile \psi$. Hence $\smile$ induces a product on cohomology classes. We write $\varphi \smile \varphi = \varphi^2$, etc.

Now suppose $A_1, A_2 \subset X$ are such that the interiors of A_1 and A_2 in $A_1 \cup A_2$ cover the union. Then an element of $C^{p+q}(X)$ that is locally zero on both A_1 and A_2 is locally zero on $A_1 \cup A_2$ also, so we get a cup product $\smile : C^p(X, A_1) \times C^q(X, A_2) \to C^{p+q}(X, A_1 \cup A_2)$, which then induces a cohomology product $\smile : H^p(X, A_1) \times H^q(X, A_2) \to H^{p+q}(X, A_1 \cup A_2)$.

Example 2.6. $H^*(\mathbb{R}\mathrm{P}^\infty) = \mathbb{Z}_2[\omega]$, the polynomial ring on a single generator $\omega \in H^1(\mathbb{R}\mathrm{P}^\infty)$, and $H^*(\mathbb{R}\mathrm{P}^{n-1}) = \mathbb{Z}_2[\omega]/(\omega^n)$, the quotient of $\mathbb{Z}_2[\omega]$ by the ideal generated by ω^n.

A neighborhood of a pair (A, B) in X is a pair (U, V) in X such that U is a neighborhood of A and V is a neighborhood of B. The set Λ of all neighborhoods of (A, B) is partially ordered downward by inclusion: $(U_\lambda, V_\lambda) \leq (U_\mu, V_\mu)$ if $i_{\lambda\mu} : (U_\mu, V_\mu) \subset (U_\lambda, V_\lambda)$. Since $(U_\lambda \cap U_\mu, V_\lambda \cap V_\mu) \subset (U_\lambda, V_\lambda), (U_\mu, V_\mu)$, we see that Λ is directed. The collection $\{H^q(U_\lambda, V_\lambda)\}_{\lambda\in\Lambda}$ of cohomology groups and the induced maps $i^*_{\lambda\mu} : H^q(U_\lambda, V_\lambda) \to H^q(U_\mu, V_\mu)$ is a directed system since $i^*_{\lambda\lambda} = id^*_{(U_\lambda,V_\lambda)} = id_{\,H^q(U_\lambda,V_\lambda)}$ and $i^*_{\lambda\nu} = (i_{\lambda\mu} i_{\mu\nu})^* = i^*_{\mu\nu} i^*_{\lambda\mu}$. The maps $j^*_\lambda : H^q(U_\lambda, V_\lambda) \to H^q(A, B)$ induced by the inclusions $j_\lambda : (A, B) \subset (U_\lambda, V_\lambda)$ satisfy $j^*_\mu i^*_{\lambda\mu} = (i_{\lambda\mu} j_\mu)^* = j^*_\lambda$, so their limit

$$j^* = \varinjlim_{\Lambda} j^*_\lambda : \varinjlim_{\Lambda} H^q(U_\lambda, V_\lambda) \to H^q(A, B)$$

is defined. Recall that X is paracompact if it is Hausdorff and every open cover has an open locally finite refinement. The proof of the following continuity property of the Alexander-Spanier cohomology theory in paracompact spaces can be found in Spanier [**126**].

Proposition 2.7. *If X is paracompact and A and B are closed, then j^* is an isomorphism for all q.*

Recall that X is locally contractible if for every $x \in X$ and neighborhood U of x, there is a neighborhood $V \subset U$ of x that is contractible in U. If X is paracompact and locally contractible, then there is an isomorphism μ from $H^q(X)$ to the singular cohomology group $H^q_s(X)$ with $\mathbb{Z}_2$-coefficients (see, e.g., Spanier [**126**]).

Finally we recall that there is also an isomorphism h from $H^q_s(X)$ to the vector space $\mathrm{Hom}(H_q(X), \mathbb{Z}_2)$ of linear maps from the singular homology group $H_q(X)$ to $\mathbb{Z}_2$ since our coefficient group $\mathbb{Z}_2$ is a field.

2.4. Principal $\mathbb{Z}_2$-Bundles

We briefly recall the notions of paracompact $\mathbb{Z}_2$-spaces, principal $\mathbb{Z}_2$-bundles, and classifying maps. A basic reference is Steenrod [**127**].

Writing the group $\mathbb{Z}_2$ multiplicatively as $\{1, -1\}$, a paracompact $\mathbb{Z}_2$-space is a paracompact space X together with a mapping $\mu : \mathbb{Z}_2 \times X \to X$, called a $\mathbb{Z}_2$-action on X, such that

$$\mu(1, x) = x, \quad -(-x) = x \quad \forall x \in X$$

where $-x := \mu(-1, x)$. The action is fixed-point free if

$$-x \neq x \quad \forall x \in X.$$

A subset A of a paracompact $\mathbb{Z}_2$-space X is invariant if

$$-A := \{\, -x : x \in A \} = A,$$

and a map $f : X \to X'$ between two such spaces is equivariant if

$$f(-x) = -f(x) \quad \forall x \in X.$$

Two spaces X and X' are equivalent if there is an equivariant homeomorphism $f : X \to X'$. We denote by $\mathcal{F}$ the set of paracompact free $\mathbb{Z}_2$-spaces, identifying equivalent spaces.

Example 2.8. Symmetric subsets of normed linear spaces that do not contain the origin are in $\mathcal{F}$, and odd maps between them are equivariant. In particular, $S^{n-1} \in \mathcal{F}$ and the antipodal map $S^{n-1} \to S^{n-1}$, $x \mapsto -x$ is equivariant.

A principal $\mathbb{Z}_2$-bundle with paracompact base is a triple $\xi = (E, p, B)$ consisting of an $E \in \mathcal{F}$, called the total space, a paracompact space B, called the base space, and a map $p : E \to B$, called the bundle projection, such that there are

(i) an open covering $\{U_\lambda\}_{\lambda \in \Lambda}$ of B,
(ii) for each $\lambda \in \Lambda$, a homeomorphism $\varphi_\lambda : U_\lambda \times \mathbb{Z}_2 \to p^{-1}(U_\lambda)$ satisfying

$$\varphi_\lambda(b, -1) = -\varphi_\lambda(b, 1), \quad p\,\varphi_\lambda(b, \pm 1) = b \quad \forall b \in B.$$

Then each $p^{-1}(b)$, called a fiber, is some pair $\{e, -e\}$, $e \in E$.

A bundle map $f : \xi \to \xi'$ consists of an equivariant map $f : E \to E'$ and a map $\overline{f} : B \to B'$ such that $p'f = \overline{f}p$, i.e., the diagram

$$\begin{array}{ccc} E & \xrightarrow{f} & E' \\ {\scriptstyle p}\downarrow & & \downarrow{\scriptstyle p'} \\ B & \xrightarrow{\overline{f}} & B' \end{array}$$

commutes. Two bundles ξ and ξ' are equivalent if there are bundle maps $f : \xi \to \xi'$ and $f' : \xi' \to \xi$ such that $f'f$ and ff' are the identity bundle maps on ξ and ξ', respectively. We denote by $\operatorname{Prin}_{\mathbb{Z}_2} B$ the set of principal $\mathbb{Z}_2$-bundles over B and $\operatorname{Prin} \mathbb{Z}_2$ the set of all principal $\mathbb{Z}_2$-bundles with paracompact base, identifying equivalent bundles.

Each paracompact free $\mathbb{Z}_2$-space can now be identified with a principal $\mathbb{Z}_2$-bundle with paracompact base as follows. Let $\overline{X} = X/\mathbb{Z}_2$ be the quotient space of $X \in \mathcal{F}$ with each x and $-x$ identified, called the orbit space of X, and $\pi : X \to \overline{X}$ the quotient map. Thcn

$$\mathcal{P} : \mathcal{F} \to \operatorname{Prin} \mathbb{Z}_2, \quad X \mapsto (X, \pi, \overline{X})$$

is a one-to-one correspondence identifying $\mathcal{F}$ with $\operatorname{Prin} \mathbb{Z}_2$.

Example 2.9. $\mathcal{P}(S^{n-1}) = (S^{n-1}, \pi, \mathbb{R}\mathrm{P}^{n-1})$ where $\pi : S^{n-1} \to \mathbb{R}\mathrm{P}^{n-1}$ identifies antipodal points.

A map $f : B \to B'$ induces a bundle $f^* \xi' = (f^*(E'), p, B) \in \operatorname{Prin} \mathbb{Z}_2$, called the pullback, where

$$f^*(E') = \{(b, e') \in B \times E' : f(b) = p'(e')\}, \quad -(b, e') = (b, -e')$$

and

$$p(b, e') = b.$$

Then $\widetilde{f} : f^*(E') \to E'$, $(b, e') \mapsto e'$ and f constitute a bundle map $\widetilde{f} : f^* \xi' \to \xi'$, i.e., the diagram

$$\begin{array}{ccc} f^*(E') & \xrightarrow{\widetilde{f}} & E' \\ {\scriptstyle p}\downarrow & & {\scriptstyle p'}\downarrow \\ B & \xrightarrow{f} & B' \end{array}$$

commutes. Homotopic maps induce equivalent bundles.

Principal $\mathbb{Z}_2$-bundles over a given paracompact space B, and hence also paracompact free $\mathbb{Z}_2$-spaces via $\mathcal{P}$, can now be classified as follows. For each $B' \in \operatorname{Prin} \mathbb{Z}_2$, we have the mapping

$$\mathcal{T} : [B, B'] \to \operatorname{Prin}_{\mathbb{Z}_2} B, \quad [f] \mapsto f^* \xi'$$

where $[B, B']$ is the set of homotopy classes of maps from B to B'.

Proposition 2.10 (Dold [**45**]). *For the bundle $\xi' = (S^\infty, \pi, \mathbb{R}P^\infty)$, called the universal principal $\mathbb{Z}_2$-bundle, $\mathcal{T}$ is a one-to-one correspondence.*

So all principal $\mathbb{Z}_2$-bundles with paracompact base are obtainable as pullbacks of the universal bundle $(S^\infty, \pi, \mathbb{R}\mathrm{P}^\infty)$, and principal $\mathbb{Z}_2$-bundles over B are classified by homotopy classes of maps from B to the base space $\mathbb{R}\mathrm{P}^\infty$ of the universal bundle, called the classifying space. Thus for each $X \in \mathcal{F}$, there is a map $f : \overline{X} \to \mathbb{R}\mathrm{P}^\infty$, unique up to homotopy and called the classifying map, such that

$$\mathcal{T}([f]) = \mathcal{P}(X).$$

2.5. Cohomological Index

We recall the construction and some properties of the cohomological index of Fadell and Rabinowitz [**49**] for $\mathbb{Z}_2$-actions. This index is defined for each paracompact $\mathbb{Z}_2$-space, but we will not have occasion to use non-free actions in this text, and therefore will define the index only for free spaces.

Let $f : \overline{X} \to \mathbb{R}\mathrm{P}^\infty$ be the classifying map of $X \in \mathcal{F}$ and

$$f^* : H^*(\mathbb{R}\mathrm{P}^\infty) \to H^*(\overline{X})$$

the induced homomorphism of the cohomology rings. Referring to Example 2.6, the $\mathbb{Z}_2$-cohomological index of X is defined by

$$i(X) = \sup \{k \geqslant 1 : f^*(\omega^{k-1}) \neq 0\},$$

which is well-defined since homotopic maps induce the same homomorphism by (c_1). Taking $\omega^0 = 1 \in H^0(\mathbb{R}\mathrm{P}^\infty)$, $f^*(\omega^0) = 1 \in H^0(\overline{X})$ and hence $i(X) \geqslant 1$ when $X \neq \varnothing$. We set $i(\varnothing) = 0$.

Example 2.11. The classifying map of S^{n-1} is the inclusion $\mathbb{R}\mathrm{P}^{n-1} \subset \mathbb{R}\mathrm{P}^\infty$, which induces isomorphisms on H^q for $q \leqslant n-1$, so $i(S^{n-1}) = n$. In particular, we have the normalization $i(\mathbb{Z}_2) = i(S^0) = 1$, and $i(S^\infty) = \infty$.

The following proposition lists the basic properties of the index.

Proposition 2.12 (Fadell-Rabinowitz [**49**]). *The index* $i : \mathcal{F} \to \mathbb{N} \cup \{0, \infty\}$ *has the following properties.*

(i_1) *Definiteness:* $i(X) = 0$ *if and only if* $X = \varnothing$.

(i_2) *Monotonicity: If* $f : X \to Y$ *is an equivariant map (in particular, if* $X \subset Y$*), then*

$$i(X) \leqslant i(Y).$$

Thus, equality holds when f *is an equivariant homeomorphism.*

(i_3) *Dimension: If* X *is a symmetric subset of a normed linear space* W *and does not contain the origin, then*

$$i(X) \leqslant \dim W.$$

(i_4) *Continuity: If* $X \in \mathcal{F}$ *and* A *is a closed invariant subset of* X*, then there is a closed invariant neighborhood* N *of* A *in* X *such that*

$$i(N) = i(A).$$

When X *is a metric space and* A *is compact,* N *may be chosen to be a* δ*-neighborhood* $N_\delta(A)$.

(i_5) *Subadditivity: If* $X \in \mathcal{F}$ *and* A *and* B *are closed invariant subsets of* X *such that* $X = A \cup B$*, then*

$$i(A \cup B) \leqslant i(A) + i(B).$$

(i_6) *Stability: If* SX *is the suspension of* $X \neq \varnothing$*, then*

$$i(SX) = i(X) + 1.$$

(i_7) *Piercing property: If* $X, A \in \mathcal{F}$, X_0 *and* X_1 *are closed invariant subsets of* X *such that* $X = X_0 \cup X_1$*, and* $\varphi : A \times [0,1] \to X$ *is a continuous mapping satisfying* $\varphi(-x,t) = -\varphi(x,t)$ *for all* (x,t), $\varphi(A \times [0,1])$ *is closed in* X, $\varphi(A \times \{0\}) \subset X_0$*, and* $\varphi(A \times \{1\}) \subset X_1$*, then*

$$i(\varphi(A \times [0,1]) \cap X_0 \cap X_1) \geqslant i(A).$$

(i_8) *Neighborhood of zero: If* U *is a bounded closed symmetric neighborhood of the origin in a normed linear space* W*, then*

$$i(\partial U) = \dim W.$$

PROOF. (i_1) This is clear from the definition of the index.

(i_2) We have the commutative diagram

$$\begin{array}{ccc} H^{k-1}(\mathbb{R}\mathrm{P}^\infty) & \xrightarrow{\pi^*} & H^{k-1}(S^\infty) \\ {\scriptstyle g^*}\downarrow & & \downarrow{\scriptstyle \tilde{g}^*} \\ H^{k-1}(\overline{Y}) & \xrightarrow{\pi^*} & H^{k-1}(Y) \\ {\scriptstyle \overline{f}^*}\downarrow & & \downarrow{\scriptstyle f^*} \\ H^{k-1}(\overline{X}) & \xrightarrow{\pi^*} & H^{k-1}(X) \end{array}$$

where g is the classifying map of Y and $\overline{f} : \overline{X} \to \overline{Y}$ is the induced map, with $g\overline{f}$ serving as a classifying map for X. Since $\overline{f}^* g^*(\omega^{k-1}) \neq 0$ implies $g^*(\omega^{k-1}) \neq 0$, the conclusion follows.

(i_3) The radial projection onto the unit sphere,

$$f : X \to S, \quad x \mapsto \frac{x}{\|x\|},$$

is odd, so

$$i(X) \leqslant i(S) = \dim W$$

by (i_2) and Example 2.11.

(i_4) If $k := i(A) = \infty$, then by (i_2), we can take $N = X$, or $N_\delta(A)$ with any $\delta > 0$ when X is a metric space, so we assume that $k < \infty$.

The set Λ of all neighborhoods of A in X is directed downward by inclusion. If U is a neighborhood of A, there is a closed neighborhood $V \subset U$ of A since X is normal and A is closed, and then $V \cap -V \subset U$ is a closed invariant neighborhood of A since A is invariant, so the subset $\mathcal{N}$ of closed invariant neighborhoods N_λ is cofinal in Λ. Thus, by Proposition 2.7,

$$\overline{j}^* = \varinjlim_{\mathcal{N}} \overline{j}_\lambda^* : \varinjlim_{\mathcal{N}} H^k(\overline{N}_\lambda) \approx H^k(\overline{A})$$

where $j_\lambda : A \subset N_\lambda$. Now, we have the commutative diagram

$$\begin{array}{ccccc} H^k(\overline{N}_\lambda) & & \xrightarrow{\overline{i}_\lambda^*} & & \varinjlim_{\mathcal{N}} H^k(\overline{N}_\lambda) \\ {\scriptstyle f_\lambda^*}\uparrow & & \searrow{\scriptstyle \overline{j}_\lambda^*} & & \downarrow{\scriptstyle \overline{j}^*}\ \approx \\ H^k(\mathbb{R}\mathrm{P}^\infty) & & \xrightarrow{(f_\lambda \overline{j}_\lambda)^*} & & H^k(\overline{A}) \end{array}$$

where f_λ is the classifying map of N_λ, with $f_\lambda \overline{j}_\lambda$ serving as a classifying map for A. Since $i(A) = k$,

$$\overline{j}^* \overline{i}_\lambda^* f_\lambda^*(\omega^k) = (f_\lambda \overline{j}_\lambda)^*(\omega^k) = 0,$$

and since $\overline{j}^*$ is an isomorphism, then $\overline{i}_\lambda^* f_\lambda^*(\omega^k) = 0$. Proposition 2.2 now gives a $\mu \in \mathcal{N}$ such that

$$(f_\lambda \overline{i}_{\lambda\mu})^*(\omega^k) = \overline{i}_{\lambda\mu}^* f_\lambda^*(\omega^k) = 0$$

where $i_{\lambda\mu} : N_\mu \subset N_\lambda$, with $f_\lambda \bar{i}_{\lambda\mu}$ serving as a classifying map for N_μ, so $i(N_\mu) \leqslant k$. The desired conclusion follows since $k \leqslant i(N_\mu)$ by (i_2).

When X is a metric space and A is compact, $\delta := \operatorname{dist}(A, N_\mu^c) > 0$, and $i(N_\delta(A)) = k$ since $A \subset N_\delta(A) \subset N_\mu$.

(i_5) Clearly, we may assume that $k := i(A)$ and $l := i(B)$ are finite. By (i_4), there are closed invariant neighborhoods M of A and N of B in X such that

$$i(M) = k, \qquad i(N) = l,$$

and we have the commutative diagram

$$\begin{array}{ccccc} H^k(\overline{X}, \overline{M}) & \xrightarrow{j_1^*} & H^k(\overline{X}) & \xrightarrow{i_1^*} & H^k(\overline{M}) \\ & & f^* \big\uparrow & \nearrow (fi_1)^* & \\ & & H^k(\mathbb{RP}^\infty) & & \end{array}$$

where the top row comes from the exact sequence of the pair $(\overline{X}, \overline{M})$ and f is the classifying map of X, with fi_1 serving as a classifying map for M. Since $i(M) = k$,

$$i_1^* f^*(\omega^k) = (fi_1)^*(\omega^k) = 0,$$

so $f^*(\omega^k) \in \ker i_1^* = j_1^*(H^k(\overline{X}, \overline{M}))$ by exactness, say, $f^*(\omega^k) = j_1^* \alpha_1$. Similarly, $f^*(\omega^l) = j_2^* \alpha_2$ for some $\alpha_2 \in H^l(\overline{X}, \overline{N})$ where $j_2 : \overline{X} \subset (\overline{X}, \overline{N})$. Now, since the interiors of M and N cover X, we have the commutative diagram

$$\begin{array}{ccc} H^k(\overline{X}, \overline{M}) \times H^l(\overline{X}, \overline{N}) & \xrightarrow{\smile} & H^{k+l}(\overline{X}, \overline{M} \cup \overline{N}) \\ j_1^* \times j_2^* \big\downarrow & & j^* \big\downarrow \\ H^k(\overline{X}) \times H^l(\overline{X}) & \xrightarrow{\smile} & H^{k+l}(\overline{X}) \end{array}$$

where the horizontal maps are cup products and $j : \overline{X} \subset (\overline{X}, \overline{M} \cup \overline{N})$. Since $\overline{M} \cup \overline{N} = \overline{X}$ and hence $H^*(\overline{X}, \overline{M} \cup \overline{N}) = 0$,

$$\begin{aligned} f^*(\omega^{k+l}) = f^*(\omega^k \smile \omega^l) = f^*(\omega^k) \smile f^*(\omega^l) &= j_1^* \alpha_1 \smile j_2^* \alpha_2 \\ &= j^*(\alpha_1 \smile \alpha_2) = j^* 0 = 0, \end{aligned}$$

so $i(X) \leqslant k + l$.

(i_6) Identifying X with $X \times \{0\} \subset SX$ via the imbedding $i_0 : X \to SX$, $x \mapsto (x, 0)$, we have $k := i(X) \leqslant i(SX)$ by (i_2), in particular, we may assume that $k < \infty$. Let $A = X \times [-1/2, 1/2]$, $B^+ = (X \times [1/2, 1])/(X \times \{1\})$, $B^- = (X \times [-1, -1/2])/(X \times \{-1\})$, and $B = B^+ \cup B^-$. Then A and B are closed invariant subsets of SX such that $SX = A \cup B$, so $i(SX) \leqslant i(A) + i(B)$ by (i_5). Since the projection onto the first factor $\mathrm{Pr}_1 : A \to X$, $(x, t) \mapsto x$ is equivariant, $i(A) \leqslant k$, and $i(B) = 1$ by Proposition 2.14 (i) below (whose proof uses only (i_2)), so $i(SX) \leqslant k + 1$. Thus,

$$k = i(X) \leqslant i(SX) \leqslant k + 1.$$

If $i(SX) = k$, then by Proposition 2.14 (iv) below (whose proof is independent of (i_6)), the rank of $i_0^* : H^{k-1}(SX) \to H^{k-1}(X)$ is at least $1 + \delta_{k1}$, contradicting the fact that i_0 is homotopic to the constant map $i_1 : X \to SX$, $x \mapsto (x, 1)$ via the imbedding $F : X \times [0, 1] \to SX$, $(x, t) \mapsto (x, t)$ and hence $\operatorname{rank} i_0^* = \operatorname{rank} i_1^* = \delta_{k1}$. So $i(SX) = k + 1$.

(i_7) Let $Y_0 = \varphi^{-1}(X_0)$, $Y_1 = \varphi^{-1}(X_1)$, and $Y = A \times [0, 1]$. Then Y_0 and Y_1 are closed invariant subsets of Y such that $Y = Y_0 \cup Y_1$. Since φ maps $B = Y_0 \cap Y_1$ into $C = \varphi(Y) \cap X_0 \cap X_1$, (i_2) gives $i(C) \geqslant i(B)$. We will show that the map $f = \mathrm{Pr}_1|_B : B \to A$, where $\mathrm{Pr}_1 : Y \to A$ is the projection onto the first factor, induces injections $\overline{f}^* : H^q(\overline{A}) \to H^q(\overline{B})$ for all q and hence $i(B) \geqslant i(A)$. Since the retraction $r = \mathrm{Pr}_1|_{Y_0} : Y_0 \to A$ induces injections $\overline{r}^*$, it suffices to show that $i_0 : B \subset Y_0$ induces injections $\overline{i}_0^*$.

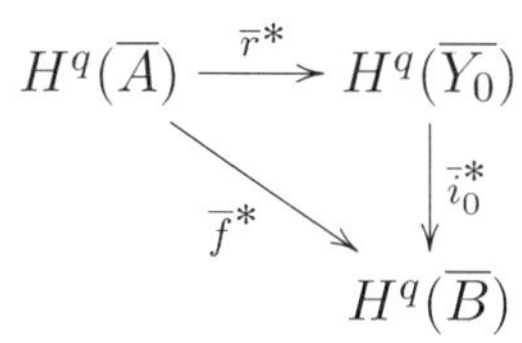

We have the inclusions

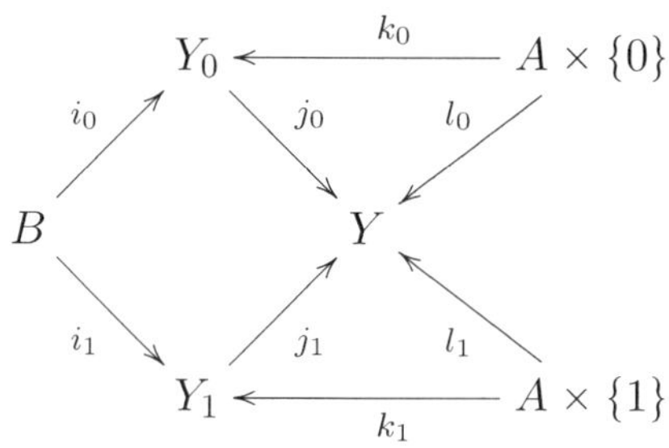

and the Mayer-Vietoris exact sequence

$$\cdots \longrightarrow H^q(\overline{Y}) \xrightarrow{j^*} H^q(\overline{Y_0}) \oplus H^q(\overline{Y_1}) \xrightarrow{i^*} H^q(\overline{B}) \longrightarrow \cdots$$

where $j^* = (\overline{j}_0^*, -\overline{j}_1^*)$ and $i^* = \overline{i}_0^* + \overline{i}_1^*$ (see, e.g., Spanier [**126**]). If $\overline{i}_0^*\alpha = 0$,

$$i^*(\alpha, 0) = \overline{i}_0^*\alpha + \overline{i}_1^* 0 = 0,$$

so $(\alpha, 0) \in \ker i^* = j^*(H^q(\overline{Y}))$ by exactness, say,

$$(\alpha, 0) = j^*\beta = (\overline{j}_0^*\beta, -\overline{j}_1^*\beta).$$

Then $\overline{j}_1^*\beta = 0$ and hence $\overline{l}_1^*\beta = \overline{k}_1^*\overline{j}_1^*\beta = 0$, which implies that $\beta = 0$ since l_1, and hence also $\overline{l}_1$, is a homotopy equivalence, so $\alpha = \overline{j}_0^*\beta = 0$.

(i_8) Taking $R > 0$ so large that $B_R(0) \supset U$ and applying Corollary 2.13 below (whose proof uses only (i_2) and (i_7)) to the identity map on $IS_R(0) = B_R(0)$ gives

$$i(\partial U) = i(B_R(0) \cap \partial U) \geqslant i(S_R(0)) = i(S) = \dim W,$$

and (i_3) gives the opposite inequality. $\square$

Corollary 2.13. *If U is a bounded closed symmetric neighborhood of the origin in a normed linear space W, A is a bounded symmetric subset of $\overline{U^c}$, and $\psi : IA \to W$ is an odd continuous mapping such that $\psi(IA)$ is closed and $\psi|_A = id_A$, then*

$$i(\psi(IA) \cap \partial U) \geqslant i(A).$$

PROOF. We have $\psi(IA) \cap \partial U = \psi(\psi^{-1}(\partial U))$, so

$$i(\psi(IA) \cap \partial U) \geqslant i(\psi^{-1}(\partial U))$$

by (i_2). Since ψ is odd, $\psi(0) = 0$, so there is a $\delta > 0$ such that

$$V_\delta := \{tx : x \in A,\ t \in [0, \delta)\} \subset \psi^{-1}(\mathring{U}) \tag{2.3}$$

by continuity. We apply (i_7) to

$$\varphi : A \times [0,1] \to IA \backslash V_\delta = (\psi^{-1}(U) \backslash V_\delta) \cup \psi^{-1}(\overline{U^c}),$$

$$(x,t) \mapsto ((1-t)\,\delta + t)\,x.$$

Since V_δ is contained in the closed set $\psi^{-1}(U)$, so is its relative boundary ∂V_δ, so $\varphi(A \times \{0\}) = \partial V_\delta \subset \psi^{-1}(U) \backslash V_\delta$. Since ψ is the identity on $A \subset \overline{U^c}$, $\varphi(A \times \{1\}) = A \subset \psi^{-1}(\overline{U^c})$. Since φ is onto,

$$\varphi(A \times [0,1]) \cap (\psi^{-1}(U) \backslash V_\delta) \cap \psi^{-1}(\overline{U^c}) = \psi^{-1}(\partial U) \backslash V_\delta = \psi^{-1}(\partial U)$$

by (2.3), so we have

$$i(\psi^{-1}(\partial U)) \geqslant i(A). \quad \square$$

The following proposition gives some additional properties of the index. Part *(iii)* is due to Degiovanni and Lancelotti [**43**], and *(iv)* to Cingolani and Degiovanni [**30**].

Proposition 2.14. *Let $X \in \mathcal{F}$ with index $k \geqslant 1$.*

- *(i) If X is the disjoint union of a pair of subsets $U, -U$, then $k = 1$. In particular, $k = 1$ when X is a finite set.*
- *(ii) If X is compact, then $k < \infty$.*
- *(iii) If X is locally contractible, then for each finite $j \leqslant k$, X has a compact invariant subset C with $i(C) \geqslant j$. In particular, there is a compact invariant subset C with index k when $k < \infty$.*
- *(iv) If $k < \infty$ and A is an invariant subset of X with index k, then the rank of $i^* : H^{k-1}(X) \to H^{k-1}(A)$, induced by $i : A \subset X$, is at least $1 + \delta_{k1}$. In particular, $\widetilde{H}^{k-1}(X) \neq 0$.*

PROOF. *(i)* The map

$$f : X \to S^0, \quad x \mapsto \begin{cases} 1, & x \in U \\ -1, & x \in -U \end{cases}$$

is equivariant, so

$$1 \leqslant k \leqslant i(S^0) = 1$$

by (i_2) and Example 2.11.

(*ii*) Each $x \in X$ has a closed neighborhood U_x such that $U_x \cap -U_x = \varnothing$ since $x \neq -x$ and X is regular, and a finite number of them, $U_{x_1}, \ldots, U_{x_m}$, cover X by compactness. Then

$$i(X) \leqslant \sum_{j=1}^{m} i(U_{x_j} \cup -U_{x_j}) = m$$

by (i_5) and (i).

(*iii*) For any invariant subset C of X, we have the commutative diagram

$$\begin{array}{ccccc}
H^{j-1}(\mathbb{R}\mathrm{P}^\infty) & \underset{\approx}{\xrightarrow{\mu}} & H_s^{j-1}(\mathbb{R}\mathrm{P}^\infty) & \underset{\approx}{\xrightarrow{h}} & \mathrm{Hom}(H_{j-1}(\mathbb{R}\mathrm{P}^\infty), \mathbb{Z}_2) \\
{\scriptstyle f^*}\downarrow & & {\scriptstyle f_s^*}\downarrow & & {\scriptstyle (f_*)^*}\downarrow \\
H^{j-1}(\overline{X}) & \underset{\approx}{\xrightarrow{\mu}} & H_s^{j-1}(\overline{X}) & \underset{\approx}{\xrightarrow{h}} & \mathrm{Hom}(H_{j-1}(\overline{X}), \mathbb{Z}_2) \\
{\scriptstyle \bar{i}^*}\downarrow & & {\scriptstyle \bar{i}_s^*}\downarrow & & {\scriptstyle (\bar{i}_*)^*}\downarrow \\
H^{j-1}(\overline{C}) & \xrightarrow{\mu} & H_s^{j-1}(\overline{C}) & \underset{\approx}{\xrightarrow{h}} & \mathrm{Hom}(H_{j-1}(\overline{C}), \mathbb{Z}_2)
\end{array}$$

where f is the classifying map of X and $\bar{i} : \overline{C} \subset \overline{X}$ is induced by $i : C \subset X$, with $f\bar{i}$ serving as a classifying map for C. Starting from the top left-hand corner, $f^*(\omega^{j-1}) \neq 0$ since $j \leqslant k$, μ is an isomorphism since $\overline{X}$ is locally contractible, and h is also an isomorphism, so the map

$$\alpha := h\,\mu f^*(\omega^{j-1}) \neq 0$$

in $\mathrm{Hom}(H_{j-1}(\overline{X}), \mathbb{Z}_2)$, say, $\alpha(z) \neq 0$, $z \in H_{j-1}(\overline{X})$. Since singular homology is a theory with compact supports, X has a compact invariant subset C such that z is in the image of $\bar{i}_* : H_{j-1}(\overline{C}) \to H_{j-1}(\overline{X})$, say, $z = \bar{i}_* c$, $c \in H_{j-1}(\overline{C})$. Then from the lower part of the diagram

$$(h\,\mu\,\bar{i}^* f^*(\omega^{j-1}))(c) = ((\bar{i}_*)^* \alpha)(c) = \alpha(\bar{i}_* c) = \alpha(z) \neq 0,$$

and hence $(f\bar{i})^*(\omega^{j-1}) = \bar{i}^* f^*(\omega^{j-1}) \neq 0$, so $i(C) \geqslant j$.

(*iv*) We have the commutative diagram

$$\begin{array}{ccccccc}
H^{k-1}(\mathbb{R}\mathrm{P}^\infty) & \xrightarrow{\pi^*} & H^{k-1}(S^\infty) & \xrightarrow{r^*} & H^{k-1}(\mathbb{R}\mathrm{P}^\infty) & \xrightarrow{\smile \omega} & H^k(\mathbb{R}\mathrm{P}^\infty) \\
{\scriptstyle f^*}\downarrow & & {\scriptstyle \tilde{f}^*}\downarrow & & {\scriptstyle f^*}\downarrow & & {\scriptstyle f^*}\downarrow \\
H^{k-1}(\overline{X}) & \xrightarrow{\pi^*} & H^{k-1}(X) & \xrightarrow{r^*} & H^{k-1}(\overline{X}) & \xrightarrow{\smile f^*\omega} & H^k(\overline{X}) \\
{\scriptstyle \bar{i}^*}\downarrow & & {\scriptstyle i^*}\downarrow & & {\scriptstyle \bar{i}^*}\downarrow & & {\scriptstyle \bar{i}^*}\downarrow \\
H^{k-1}(\overline{A}) & \xrightarrow{\pi^*} & H^{k-1}(A) & \xrightarrow{r^*} & H^{k-1}(\overline{A}) & \xrightarrow{\smile \bar{i}^* f^*\omega} & H^k(\overline{A})
\end{array}$$

where f is the classifying map of X, $\bar{i} : \overline{A} \subset \overline{X}$ is induced by $i : A \subset X$, with $f\bar{i}$ serving as a classifying map for A, and the horizontal rows come from the Thom-Gysin exact sequence (see, e.g., Spanier [**126**]). Since $i(X) = k$,

$$f^*(\omega^{k-1}) \smile f^*\omega = f^*(\omega^{k-1} \smile \omega) = f^*(\omega^k) = 0,$$

so $f^*(\omega^{k-1}) \in \ker(\smile f^*\omega) = r^*(H^{k-1}(X))$ by exactness, say, $f^*(\omega^{k-1}) = r^*\alpha$. Then from the lower middle square

$$r^*i^*\alpha = \bar{i}^*r^*\alpha = \bar{i}^*f^*(\omega^{k-1}) = (f\bar{i})^*(\omega^{k-1}) \neq 0$$

since $i(A) = k$, so $i^*\alpha \neq 0$. When $k = 1$, $i^*\alpha \neq \pi^*1 = 1$ since $i^*\alpha \notin \ker r^* = \pi^*(H^0(\overline{A}))$, and $i^*1 = 1$. The second assertion follows by taking $A = X$. □

CHAPTER 3

Critical Point Theory

In this chapter we develop the critical point theory needed for our treatment of equation (1.1).

Let Φ be a real-valued function defined on a real Banach space W of dimension $d \geqslant 1$. We say that Φ is Fréchet differentiable at $u \in W$ if there is an element $\Phi'(u)$ of the dual W^*, called the Fréchet derivative of Φ at u, such that

$$\Phi(u+v) = \Phi(u) + \langle \Phi'(u), v \rangle + \mathrm{o}(\|v\|) \quad \text{as} \quad v \to 0 \quad \text{in} \quad W$$

where $\langle \cdot, \cdot \rangle$ is the duality pairing. The functional Φ is continuously Fréchet differentiable on W, or belongs to the class $C^1(W, \mathbb{R})$, if Φ' is defined everywhere and the map $W \to W^*$, $u \mapsto \Phi'(u)$ is continuous. We assume that $\Phi \in C^1(W, \mathbb{R})$ for the rest of this chapter. Replacing Φ with $\Phi - \Phi(0)$ if necessary, we may also assume that $\Phi(0) = 0$.

We say that u is a critical point of Φ if $\Phi'(u) = 0$. A real number c is a critical value of Φ if there is a critical point u with $\Phi(u) = c$, otherwise it is a regular value (in particular, any c that is not a value of Φ is a regular value!). We use the notations

$$\Phi_a = \{u \in W : \Phi(u) \geqslant a\}, \qquad \Phi^b = \{u \in W : \Phi(u) \leqslant b\},$$

$$\Phi_a^b = \Phi_a \cap \Phi^b,$$

$$K = \{u \in W : \Phi'(u) = 0\}, \qquad \widetilde{W} = W \backslash K,$$

$$K_a^b = K \cap \Phi_a^b, \qquad K^c = K_c^c$$

for the various superlevel, sublevel, critical, and regular sets of Φ.

The functional Φ is called even if it is invariant under the antipodal action of $\mathbb{Z}_2$, i.e.,

$$\Phi(-u) = \Phi(u) \quad \forall u \in W.$$

Then Φ' is odd, i.e.,

$$\Phi'(-u) = -\Phi'(u) \quad \forall u \in W.$$

We denote by

$$S = \{u \in W : \|u\| = 1\}$$

the unit sphere in W and

$$\pi_S : W \backslash \{0\} \to S, \quad u \mapsto \frac{u}{\|u\|}$$

the radial projection onto S.

3.1. Compactness Conditions

It is usually necessary to have some "compactness" when seeking critical points of a functional. The following condition was originally introduced by Palais and Smale [**91**].

Definition 3.1. Φ satisfies the Palais-Smale compactness condition at the level c, or $(\mathrm{PS})_c$ for short, if every sequence $(u_j) \subset W$ such that

$$\Phi(u_j) \to c, \quad \Phi'(u_j) \to 0,$$

called a $(\mathrm{PS})_c$ sequence, has a convergent subsequence. Φ satisfies (PS) if it satisfies $(\mathrm{PS})_c$ for every $c \in \mathbb{R}$, or equivalently, if every sequence such that

$$(\Phi(u_j)) \text{ is bounded}, \quad \Phi'(u_j) \to 0,$$

called a (PS) sequence, has a convergent subsequence.

The following weaker version was introduced by Cerami [**25**].

Definition 3.2. Φ satisfies the Cerami condition at the level c, or $(\mathrm{C})_c$ for short, if every sequence such that

$$\Phi(u_j) \to c, \quad \big(1 + \|u_j\|\big)\, \Phi'(u_j) \to 0,$$

called a $(\mathrm{C})_c$ sequence, has a convergent subsequence. Φ satisfies (C) if it satisfies $(\mathrm{C})_c$ for every c, or equivalently, if every sequence such that

$$(\Phi(u_j)) \text{ is bounded}, \quad \big(1 + \|u_j\|\big)\, \Phi'(u_j) \to 0,$$

called a (C) sequence, has a convergent subsequence.

This condition is weaker since a $(\mathrm{C})_c$ (resp. (C)) sequence is clearly a $(\mathrm{PS})_c$ (resp. (PS)) sequence also. Note that the limit of a $(\mathrm{PS})_c$ (resp. (PS)) sequence is in K^c (resp. K) since Φ is C^1. Since any sequence in K^c is a $(\mathrm{C})_c$ sequence, it follows that K^c is compact when $(\mathrm{C})_c$ holds.

It suffices to show that (u_j) is bounded when verifying these conditions for the functional in (1.4) by

Lemma 3.3. *If Φ is as in* (1.4), *then every bounded sequence (u_j) such that $\Phi'(u_j) \to 0$ has a convergent subsequence.*

Proof. A renamed subsequence converges weakly to some u since W is reflexive, and $f(u_j)$ converges in W^* for a further subsequence since f is compact. Then

$$\langle A_p\, u_j, u_j - u\rangle = \big\langle \Phi'(u_j) + f(u_j), u_j - u\big\rangle \to 0$$

by Lemma 3.4 below, so $u_j \to u$ for a subsequence by (A_4). □

Lemma 3.4. *If $(L_j) \subset W^*$ converges to L and $(u_j) \subset W$ converges weakly to u, then $\langle L_j, u_j\rangle \to \langle L, u\rangle$, in particular, $\langle L_j, u_j - u\rangle \to 0$.*

Proof. Since (u_j) is bounded by the principle of uniform boundedness,

$$|\,\langle L_j, u_j\rangle - \langle L, u\rangle\,| \leqslant \|L_j - L\|^*\, \|u_j\| + |\,\langle L, u_j\rangle - \langle L, u\rangle\,| \to 0. \quad \square$$

3.2. Deformation Lemmas

Deformation lemmas allow to lower sublevel sets of a functional, away from its critical set, and are an essential tool for locating critical points. The main ingredient in their proofs is usually a suitable negative pseudo-gradient flow, a notion due to Palais [**93**].

Definition 3.5. A pseudo-gradient vector field for Φ on $\widetilde{W}$ is a locally Lipschitz continuous mapping $V : \widetilde{W} \to W$ satisfying

$$\|V(u)\| \leqslant \|\Phi'(u)\|^*, \quad 2\left\langle \Phi'(u), V(u)\right\rangle \geqslant \left(\|\Phi'(u)\|^*\right)^2 \quad \forall u \in \widetilde{W}. \tag{3.1}$$

Lemma 3.6. *There is a pseudo-gradient vector field V for Φ on $\widetilde{W}$. When Φ is even, V may be chosen to be odd.*

PROOF. For each $u \in \widetilde{W}$, there is a $w(u) \in W$ satisfying

$$\|w(u)\| < \|\Phi'(u)\|^*, \quad 2\left\langle \Phi'(u), w(u)\right\rangle > \left(\|\Phi'(u)\|^*\right)^2$$

by the definition of the norm in W^*. Since Φ' is continuous, then

$$\|w(u)\| \leqslant \|\Phi'(v)\|^*, \quad 2\left\langle \Phi'(v), w(u)\right\rangle \geqslant \left(\|\Phi'(v)\|^*\right)^2 \quad \forall v \in N_u \tag{3.2}$$

for some open neighborhood $N_u \subset \widetilde{W}$ of u.

Since $\widetilde{W}$ is a metric space and hence paracompact, the open covering $\{N_u\}_{u \in \widetilde{W}}$ has a locally finite refinement, i.e., an open covering $\{N_\lambda\}_{\lambda \in \Lambda}$ of $\widetilde{W}$ such that

(*i*) each $N_\lambda \subset N_{u_\lambda}$ for some $u_\lambda \in \widetilde{W}$,

(*ii*) each $u \in \widetilde{W}$ has a neighborhood U_u that intersects N_λ only for λ in some finite subset Λ_u of Λ

(see, e.g., Kelley [**59**]). Let $\{\varphi_\lambda\}_{\lambda \in \Lambda}$ be a Lipschitz continuous partition of unity subordinate to $\{N_\lambda\}_{\lambda \in \Lambda}$, i.e.,

(*i*) $\varphi_\lambda \in \operatorname{Lip}\left(\widetilde{W}, [0,1]\right)$ vanishes outside N_λ,

(*ii*) for each $u \in \widetilde{W}$,

$$\sum_{\lambda \in \Lambda} \varphi_\lambda(u) = 1 \tag{3.3}$$

where the sum is actually over a subset of Λ_u,

for example,

$$\varphi_\lambda(u) = \frac{\operatorname{dist}(u, \widetilde{W} \backslash N_\lambda)}{\sum\limits_{\lambda \in \Lambda} \operatorname{dist}(u, \widetilde{W} \backslash N_\lambda)}.$$

Now

$$V(u) = \sum_{\lambda \in \Lambda} \varphi_\lambda(u)\, w(u_\lambda)$$

is Lipschitz in each U_u and satisfies (3.1) by (3.2) and (3.3).

When Φ is even, Φ' is odd and hence $-V(-u)$ is also a pseudo-gradient, and therefore so is the odd convex combination $\dfrac{1}{2}\,(V(u) - V(-u))$. □

The following deformation lemma improves that of Cerami [**25**].

Lemma 3.7 (First Deformation Lemma). *If $c \in \mathbb{R}$, C is a bounded set containing K^c, $\delta, k > 0$, and Φ satisfies $(\mathrm{C})_c$, then there are an $\varepsilon_0 > 0$ and, for each $\varepsilon \in (0, \varepsilon_0)$, a map $\eta \in C([0,1] \times W, W)$ satisfying*

(*i*) $\eta(0, \cdot) = id_W$,
(*ii*) *$\eta(t, \cdot)$ is a homeomorphism of W for all $t \in [0,1]$,*
(*iii*) *$\eta(t, \cdot)$ is the identity outside $A = \Phi_{c-2\varepsilon}^{c+2\varepsilon} \backslash N_{\delta/3}(C)$ for all $t \in [0,1]$,*
(*iv*) $\|\eta(t,u) - u\| \leqslant \big(1 + \|u\|\big)\, \delta/k \quad \forall (t,u) \in [0,1] \times W$,
(*v*) *$\Phi(\eta(\cdot, u))$ is nonincreasing for all $u \in W$,*
(*vi*) $\eta(1, \Phi^{c+\varepsilon} \backslash N_\delta(C)) \subset \Phi^{c-\varepsilon}$.

When Φ is even and C is symmetric, η may be chosen so that $\eta(t, \cdot)$ is odd for all $t \in [0,1]$.

First we prove a lemma.

Lemma 3.8. *If $c \in \mathbb{R}$, N is an open neighborhood of K^c, $k > 0$, and Φ satisfies $(\mathrm{C})_c$, then there is an $\varepsilon_0 > 0$ such that*

$$\inf_{u \in \Phi_{c-\varepsilon}^{c+\varepsilon} \backslash N} \big(1 + \|u\|\big)\, \|\Phi'(u)\|^* \geqslant k\, \varepsilon \quad \forall \varepsilon \in (0, \varepsilon_0).$$

Proof. If not, there are sequences $\varepsilon_j \searrow 0$ and $u_j \in \Phi_{c-\varepsilon_j}^{c+\varepsilon_j} \backslash N$ such that

$$\big(1 + \|u_j\|\big)\, \|\Phi'(u_j)\|^* < k\, \varepsilon_j.$$

Then $(u_j) \subset W \backslash N$ is a $(\mathrm{C})_c$ sequence and hence has a subsequence converging to some $u \in K^c \backslash N = \varnothing$, a contradiction. □

Proof of Lemma 3.7. Taking k larger if necessary, we may assume that

$$\big(1 + \|u\|\big)/k < 1/3 \quad \forall u \in N_\delta(C). \tag{3.4}$$

By Lemma 3.8, there is an $\varepsilon_0 > 0$ such that for each $\varepsilon \in (0, \varepsilon_0)$,

$$\big(1 + \|u\|\big)\, \|\Phi'(u)\|^* \geqslant \frac{8\varepsilon}{\ln(1 + \delta/k)} \quad \forall u \in A. \tag{3.5}$$

Let V be a pseudo-gradient vector field for Φ, $g \in \mathrm{Lip}_{\mathrm{loc}}(W, [0,1])$ satisfy $g = 0$ outside A and $g = 1$ on $B = \Phi_{c-\varepsilon}^{c+\varepsilon} \backslash N_{2\delta/3}(C)$, for example,

$$g(u) = \frac{\operatorname{dist}(u, A^c)}{\operatorname{dist}(u, A^c) + \operatorname{dist}(u, B)},$$

and $\eta(t,u)$, $0 \leqslant t < T(u) \leqslant \infty$ the maximal solution of

$$\dot{\eta} = -4\varepsilon\, g(\eta)\, \frac{V(\eta)}{\|V(\eta)\|^2}, \quad t > 0, \qquad \eta(0,u) = u \in W. \tag{3.6}$$

For $0 \leqslant s < t < T(u)$,

$$\begin{aligned}
\|\eta(t,u) - \eta(s,u)\| &\leqslant 4\varepsilon \int_s^t \frac{g(\eta(\tau,u))}{\|V(\eta(\tau,u))\|}\, d\tau \\
&\leqslant 8\varepsilon \int_s^t \frac{g(\eta(\tau,u))}{\|\Phi'(\eta(\tau,u))\|^*}\, d\tau && \text{by (3.1)} \\
&\leqslant \ln(1+\delta/k) \int_s^t \big(1 + \|\eta(\tau,u)\|\,\big)\, d\tau && \text{by (3.5)} \\
&\leqslant \ln(1+\delta/k) \Bigg[\int_s^t \|\eta(\tau,u) - \eta(s,u)\|\, d\tau \\
&\qquad + \big(1 + \|\eta(s,u)\|\,\big)(t-s) \Bigg],
\end{aligned}$$

and integrating gives

$$\|\eta(t,u) - \eta(s,u)\| \leqslant \big(1 + \|\eta(s,u)\|\,\big)\big((1+\delta/k)^{t-s} - 1\big). \tag{3.7}$$

Taking $s = 0$ we see that $\|\eta(\cdot,u)\|$ is bounded if $T(u) < \infty$, so $T(u) = \infty$ and (i) - (iv) follow.

By (3.6) and (3.1),

$$\begin{aligned}
\frac{d}{dt}\big(\Phi(\eta(t,u))\big) = \big\langle \Phi'(\eta), \dot{\eta} \big\rangle &= -4\varepsilon\, g(\eta)\, \frac{\langle \Phi'(\eta), V(\eta)\rangle}{\|V(\eta)\|^2} \\
&\leqslant -2\varepsilon\, g(\eta) \leqslant 0
\end{aligned} \tag{3.8}$$

and hence (v) holds. To see that (vi) holds, let $u \in \Phi^{c+\varepsilon} \backslash N_\delta(C)$ and suppose that $\eta(1,u) \notin \Phi^{c-\varepsilon}$. Then $\eta(t,u) \in \Phi_{c-\varepsilon}^{c+\varepsilon}$ for all $t \in [0,1]$, and we claim that $\eta(t,u) \notin N_{2\delta/3}(C)$. Otherwise there are $0 < s < t \leqslant 1$ such that

$$\operatorname{dist}(\eta(s,u), C) = \delta, \qquad 2\delta/3 < \operatorname{dist}(\eta(\tau,u), C) < \delta, \quad \tau \in (s,t),$$
$$\operatorname{dist}(\eta(t,u), C) = 2\delta/3.$$

But, then

$$\delta/3 \leqslant \|\eta(t,u) - \eta(s,u)\| \leqslant \big(1 + \|\eta(s,u)\|\,\big)\, \delta/k < \delta/3$$

by (3.7) and (3.4), a contradiction. Thus, $\eta(t,u) \in B$ and hence $g(\eta(t,u)) = 1$ for all $t \in [0,1]$, so (3.8) gives

$$\Phi(\eta(1,u)) \leqslant \Phi(u) - 2\varepsilon \leqslant c - \varepsilon,$$

a contradiction.

When Φ is even and C is symmetric, A and B are symmetric and hence g is even, so η may be chosen to be odd in u by choosing V to be odd. □

Lemma 3.7 provides a local deformation near a (possibly critical) level set of a functional. The following lemma shows that the homotopy type of sublevel sets can change only when crossing a critical level.

Lemma 3.9 (Second Deformation Lemma). *If $-\infty < a < b \leqslant +\infty$ and Φ has only a finite number of critical points at the level a, has no critical values in (a, b), and satisfies $(\mathrm{C})_c$ for all $c \in [a, b] \cap \mathbb{R}$, then Φ^a is a deformation retract of $\Phi^b \backslash K^b$.*

Proof. Let V be a pseudo-gradient vector field for Φ and $\zeta(t, u)$ the solution of

$$\dot{\zeta} = -\frac{V(\zeta)}{\|V(\zeta)\|^2}, \quad t > 0, \qquad \zeta(0, u) = u \in \Phi^b \backslash (\Phi^a \cup K^b) \subset \widetilde{W}. \tag{3.9}$$

Then

$$\frac{d}{dt}\big(\Phi(\zeta(t, u))\big) = -\frac{\langle \Phi'(\zeta), V(\zeta)\rangle}{\|V(\zeta)\|^2} \leqslant -\frac{1}{2}$$

by (3.1) and hence

$$\Phi(\zeta(t, u)) \leqslant \Phi(\zeta(s, u)) - \frac{1}{2}(t - s), \quad 0 \leqslant s < t. \tag{3.10}$$

Taking $s = 0$ and using $\Phi(\zeta(0, u)) = \Phi(u) \leqslant b$ gives a $T(u) \in (0, 2(b - a)]$ such that $\Phi(\zeta(t, u)) \searrow a$ as $t \nearrow T(u)$. We set $T(u) = 0$ and $\zeta(0, u) = u$ for $u \in \Phi^a$.

For $\delta > 0$, let

$$\mathcal{T}_\delta = \big\{t \in [0, T(u)) : \operatorname{dist}(\zeta(t, u), K^a) \geqslant \delta\big\}.$$

Then

$$m_\delta := \inf_{t \in \mathcal{T}_\delta} \big(1 + \|\zeta(t, u)\|\,\big)\, \|\Phi'(\zeta(t, u))\|^* > 0$$

by (C), so

$$\begin{aligned} \|\zeta(t, u) - \zeta(s, u)\| &\leqslant \int_s^t \frac{d\tau}{\|V(\zeta(\tau, u))\|} && \text{by (3.9)} \\ &\leqslant 2\int_s^t \frac{d\tau}{\|\Phi'(\zeta(\tau, u))\|^*} && \text{by (3.1)} \\ &\leqslant \frac{2}{m_\delta}\int_s^t \big(1 + \|\zeta(\tau, u)\|\,\big)\, d\tau, && [s, t] \subset \mathcal{T}_\delta. \end{aligned} \tag{3.11}$$

Case 1: $\zeta(\cdot, u)$ is bounded away from K^a. Then $\mathcal{T}_\delta = [0, T(u))$ for some $\delta > 0$ and $\|\zeta(\cdot, u)\|$ is bounded as in the proof of Lemma 3.7, so

$$\|\zeta(t, u) - \zeta(s, u)\| \leqslant C\,(t - s), \quad 0 \leqslant s < t < T(u) \tag{3.12}$$

for some $C > 0$. Let $t_j \nearrow T(u)$. Taking $t = t_j$ and $s = t_k$ in (3.12) shows that $(\zeta(t_j, u))$ is a Cauchy sequence and hence converges to some $v \in \Phi^{-1}(a) \backslash K^a$. Now taking $s = t_j$ shows that $\zeta(t, u) \to v$ as $t \nearrow T(u)$. We set $\zeta(T(u), u) = v$ and note that T is continuous in u in this case.

Case 2: $\zeta(\cdot, u)$ is not bounded away from K^a. We claim that then $\zeta(t, u)$ converges to some $v \in K^a$ as $t \nearrow T(u)$. Since K^a is a finite set, otherwise there are a $v_0 \in K^a$, $\delta > 0$ such that the ball $B_{3\delta}(v_0)$ contains no other points of K^a, and sequences $s_j, t_j \nearrow T(u)$, $s_j < t_j$ such that

$$\|\zeta(s_j, u) - v_0\| = \delta, \qquad \delta < \|\zeta(\tau, u) - v_0\| < 2\delta, \quad \tau \in (s_j, t_j),$$
$$\|\zeta(t_j, u) - v_0\| = 2\delta.$$

But, then

$$\delta \leqslant \|\zeta(t_j, u) - \zeta(s_j, u)\| \leqslant C\,(t_j - s_j) \to 0$$

by (3.11), a contradiction.

We set $\zeta(T(u), u) = v$ and claim that T is continuous in u in this case also. To see this, suppose that $u_j \to u$. We will show that

$$\underline{T} := \varliminf T(u_j) \geqslant T(u) \geqslant \varlimsup T(u_j) =: \overline{T}$$

and hence $T(u_j) \to T(u)$. If $\underline{T} < T(u)$, then passing to a subsequence, $\zeta(T(u_j), u_j) \to \zeta(\underline{T}, u)$ and hence

$$a = \Phi(\zeta(T(u_j), u_j)) \to \Phi(\zeta(\underline{T}, u)) > a,$$

a contradiction. If $\overline{T} > T(u)$, then for a subsequence and any $t < T(u)$,

$$a = \Phi(\zeta(T(u_j), u_j)) \leqslant \Phi(\zeta(t, u_j)) - \frac{1}{2}\,(T(u_j) - t)$$

by (3.10), and passing to the limit gives

$$a \leqslant \Phi(\zeta(t, u)) - \frac{1}{2}\,(\overline{T} - t) \to a - \frac{1}{2}\,(\overline{T} - T(u)) < a \text{ as } t \nearrow T(u),$$

again a contradiction.

We will show that ζ is continuous. Then

$$\eta(u, t) = \zeta(t\,T(u), u)$$

will be a deformation retraction of $\Phi^b \backslash K^b$ onto Φ^a.

Case 1: $u \in \Phi^b \backslash (\Phi^a \cup K^b)$, $0 \leqslant t < T(u)$. Then ζ is continuous at (t, u) by standard ODE theory.

Case 2: $u \in \Phi^b \backslash (\Phi^a \cup K^b)$, $t = T(u)$. Suppose that $u_j \in \Phi^b \backslash (\Phi^a \cup K^b)$, $0 \leqslant t_j \leqslant T(u_j)$, $(t_j, u_j) \to (T(u), u)$, but $\zeta(t_j, u_j) \not\to \zeta(T(u), u) =: v$. Then there is a $\delta > 0$ such that $(B_{3\delta}(v) \backslash \{v\}) \cap K^a = \varnothing$ and

$$\|\zeta(t_j, u_j) - v\| \geqslant 2\delta \tag{3.13}$$

for a subsequence. Since $\zeta(s, u)$ converges to v as $s \nearrow T(u)$,

$$\|\zeta(s, u) - v\| \leqslant \delta/2 \tag{3.14}$$

for all $s < T(u)$ sufficiently close to $T(u)$. For each such s,

$$\|\zeta(s, u_j) - \zeta(s, u)\| \leqslant \delta/2 \tag{3.15}$$

for all sufficiently large j by Case 1. Taking a sequence $s_j \nearrow T(u)$ and combining (3.14) and (3.15) gives $t_j > s_j$ and

$$\|\zeta(s_j, u_j) - v\| \leqslant \delta \tag{3.16}$$

for a further subsequence of (t_j, u_j). By (3.13) and (3.16), there are sequences $s_j', t_j' \nearrow T(u)$, $s_j \leqslant s_j' < t_j' \leqslant t_j$ such that

$$\|\zeta(s_j', u_j) - v\| = \delta, \qquad \delta < \|\zeta(\tau, u_j) - v\| < 2\delta, \quad \tau \in (s_j', t_j'),$$
$$\|\zeta(t_j', u_j) - v\| = 2\delta. \tag{3.17}$$

Then

$$\delta \leqslant \|\zeta(t_j', u_j) - \zeta(s_j', u_j)\| \leqslant C\,(t_j' - s_j') \to 0 \tag{3.18}$$

by (3.11), a contradiction.

Case 3: $u \in \Phi^{-1}(a)$, $t = 0$. Suppose that $u_j \in \Phi^b \backslash K^b$, $0 \leqslant t_j \leqslant T(u_j)$, $(t_j, u_j) \to (0, u)$, but $\zeta(t_j, u_j) \nrightarrow \zeta(0, u) = u$. Then there is a $\delta > 0$ such that $(B_{3\delta}(u) \backslash \{u\}) \cap K^a = \varnothing$ and

$$\|\zeta(t_j, u_j) - u\| \geqslant 2\delta \tag{3.19}$$

for a subsequence. Since $\zeta(0, u_j) = u_j \to u$,

$$\|\zeta(0, u_j) - u\| \leqslant \delta \tag{3.20}$$

for sufficiently large j. By (3.19) and (3.20), $u_j \in \Phi^b \backslash (\Phi^a \cup K^b)$, $t_j > 0$, and there are sequences $0 \leqslant s_j' < t_j' \leqslant t_j$ for which (3.17), and hence also (3.18), holds.

Case 4: $u \in \Phi^a \backslash \Phi^{-1}(a)$, $t = 0$. Then $\zeta(0, u) = u$. □

3.3. Minimax Principle

First deformation lemma implies that if c is a regular value and Φ satisfies $(\mathrm{C})_c$, then the family $\mathcal{D}_{c,\varepsilon}$ of maps $\eta \in C([0,1] \times W, W)$ satisfying

(*i*) $\eta(0, \cdot) = id_W$,
(*ii*) $\eta(t, \cdot)$ is a homeomorphism of W for all $t \in [0,1]$,
(*iii*) $\eta(t, \cdot)$ is the identity outside $\Phi_{c-2\varepsilon}^{c+2\varepsilon}$ for all $t \in [0,1]$,
(*iv*) $\Phi(\eta(\cdot, u))$ is nonincreasing for all $u \in W$,
(*v*) $\eta(1, \Phi^{c+\varepsilon}) \subset \Phi^{c-\varepsilon}$

is nonempty for all sufficiently small $\varepsilon > 0$. We say that a family $\mathcal{F}$ of subsets of W is invariant under $\mathcal{D}_{c,\varepsilon}$ if

$$M \in \mathcal{F},\ \eta \in \mathcal{D}_{c,\varepsilon} \implies \eta(1, M) \in \mathcal{F}.$$

Proposition 3.10 (Minimax Principle). *If $\mathcal{F}$ is a family of subsets of W,*

$$c := \inf_{M \in \mathcal{F}} \sup_{u \in M} \Phi(u) \tag{3.21}$$

is finite, $\mathcal{F}$ is invariant under $\mathcal{D}_{c,\varepsilon}$ for all sufficiently small $\varepsilon > 0$, and Φ satisfies $(\mathrm{C})_c$, then c is a critical value of Φ.

PROOF. If not, taking $\varepsilon > 0$ sufficiently small, $M \in \mathcal{F}$ with $\sup \Phi(M) \leqslant c + \varepsilon$, and $\eta \in \mathcal{D}_{c,\varepsilon}$, we have $\eta(1, M) \in \mathcal{F}$ and $\sup \Phi(\eta(1, M)) \leqslant c - \varepsilon$ by (*v*), contradicting (3.21). □

We say that a family Γ of continuous maps γ from a topological space X into W is invariant under $\mathcal{D}_{c,\varepsilon}$ if

$$\gamma \in \Gamma,\ \eta \in \mathcal{D}_{c,\varepsilon} \implies \eta(1,\cdot) \circ \gamma \in \Gamma.$$

Minimax principle is often applied in the following form.

Proposition 3.11 (Minimax Principle). *If Γ is a family of continuous maps γ from a topological space X into W,*

$$c := \inf_{\gamma \in \Gamma} \sup_{u \in \gamma(X)} \Phi(u)$$

is finite, Γ is invariant under $\mathcal{D}_{c,\varepsilon}$ for all sufficiently small $\varepsilon > 0$, and Φ satisfies $(\mathrm{C})_c$, then c is a critical value of Φ.

PROOF. Apply Proposition 3.10 with $\mathcal{F} = \{\gamma(X) : \gamma \in \Gamma\}$. □

3.4. Critical Groups

In Morse theory the local behavior of Φ near an isolated critical point u is described by the sequence of critical groups

$$C^q(\Phi, u) := H^q(\Phi^c \cap U, \Phi^c \cap U \backslash \{u\}), \quad q \geqslant 0 \tag{3.22}$$

where $c = \Phi(u)$ is the corresponding critical value and U is a neighborhood of u containing no other critical points of Φ. They are independent of U, and hence well-defined, by (c_3). Critical groups help distinguish between different types of critical points and are extremely useful for obtaining multiple critical points of a functional (see, e.g., Chang [**28**]).

One of the consequences of the second deformation lemma is the following proposition relating the change in the topology of sublevel sets across a critical level to the critical groups of the critical points at that level.

Proposition 3.12. *If $-\infty < a < b \leqslant +\infty$ and Φ has only a finite number of critical points at the level $c \in (a, b)$, has no other critical values in $[a, b]$, and satisfies $(\mathrm{C})_{c'}$ for all $c' \in [a, b] \cap \mathbb{R}$, then*

$$H^q(\Phi^b, \Phi^a) \approx \bigoplus_{u \in K^c} C^q(\Phi, u) \quad \forall q.$$

In particular,

$$\dim H^q(\Phi^b, \Phi^a) = \sum_{u \in K^c} \dim C^q(\Phi, u) \quad \forall q.$$

PROOF. We have

$$H^q(\Phi^b, \Phi^a) \approx H^q(\Phi^c, \Phi^a) \approx H^q(\Phi^c, \Phi^c \backslash K^c) \tag{3.23}$$

since Φ^c and Φ^a are deformation retracts of Φ^b and $\Phi^c \backslash K^c$, respectively, by Lemma 3.9. Taking $\delta > 0$ so small that the balls $B_\delta(u)$, $u \in K^c$ are mutually disjoint and then excising $\Phi^c \backslash \bigcup_{u \in K^c} B_\delta(u)$, we see that the last group in (3.23) is isomorphic to

$$\bigoplus_{u \in K^c} H^q(\Phi^c \cap B_\delta(u), \Phi^c \cap B_\delta(u) \backslash \{u\}) = \bigoplus_{u \in K^c} C^q(\Phi, u). \quad \square$$

For the change in the topology across multiple critical levels, we have

Proposition 3.13. *If* $-\infty < a < b \leqslant +\infty$ *are regular values and* Φ *has only a finite number of critical points in* Φ_a^b *and satisfies* $(\mathrm{C})_c$ *for all* $c \in [a, b] \cap \mathbb{R}$*, then*

$$\dim H^q(\Phi^b, \Phi^a) \leqslant \sum_{u \in K_a^b} \dim C^q(\Phi, u) \quad \forall q.$$

In particular, Φ *has a critical point* u *with* $a < \Phi(u) < b$ *and* $C^q(\Phi, u) \neq 0$ *when* $H^q(\Phi^b, \Phi^a) \neq 0$.

First we prove a lemma of a purely topological nature.

Lemma 3.14. *If* $X_1 \subset \cdots \subset X_{k+1}$ *are topological spaces, then*

$$\dim H^q(X_{k+1}, X_1) \leqslant \sum_{i=1}^{k} \dim H^q(X_{i+1}, X_i) \quad \forall q. \tag{3.24}$$

PROOF. In the exact sequence

$$\cdots \longrightarrow H^q(X_{k+1}, X_k) \xrightarrow{j^*} H^q(X_{k+1}, X_1) \xrightarrow{i^*} H^q(X_k, X_1) \longrightarrow \cdots$$

of the triple (X_{k+1}, X_k, X_1),

$$\operatorname{im} i^* \approx H^q(X_{k+1}, X_1)/\ker i^* = H^q(X_{k+1}, X_1)/\operatorname{im} j^*$$

and hence

$$\begin{aligned} \dim H^q(X_{k+1}, X_1) &= \operatorname{rank} i^* + \operatorname{rank} j^* \\ &\leqslant \dim H^q(X_k, X_1) + \dim H^q(X_{k+1}, X_k). \end{aligned}$$

Since equality holds in (3.24) when $k = 1$, the conclusion now follows by induction on k. □

PROOF OF PROPOSITION 3.13. Let $c_1 < \cdots < c_k$ be the critical values in (a, b) and

$$a = a_1 < c_1 < a_2 < c_2 < \cdots < c_{k-1} < a_k < c_k < a_{k+1} = b.$$

Applying Lemma 3.14 with $X_i = \Phi^{a_i}$ and using Proposition 3.12 gives

$$\begin{aligned} \dim H^q(\Phi^b, \Phi^a) &\leqslant \sum_{i=1}^{k} \dim H^q(\Phi^{a_{i+1}}, \Phi^{a_i}) \\ &= \sum_{i=1}^{k} \sum_{u \in K^{c_i}} \dim C^q(\Phi, u) \\ &= \sum_{u \in K_a^b} \dim C^q(\Phi, u). \quad \square \end{aligned}$$

When the critical values are bounded from below and Φ satisfies (C), the global behavior of Φ can be described by the critical groups at infinity

$$C^q(\Phi, \infty) := H^q(W, \Phi^a), \quad q \geqslant 0 \tag{3.25}$$

where a is less than all critical values. They are independent of a, and hence well-defined, by Lemma 3.9 and the homotopy invariance of the cohomology groups. Since W is contractible and $\Phi^a = \varnothing$ if and only if Φ is bounded from below, the following proposition is a consequence of Proposition 2.4.

Proposition 3.15. *Assume that Φ satisfies* (C)*.*

(*i*) *If Φ is bounded from below, then*

$$C^q(\Phi, \infty) \approx \delta_{q0}\, \mathbb{Z}_2.$$

(*ii*) *If Φ is unbounded from below, then*

$$C^q(\Phi, \infty) \approx \widetilde{H}^{q-1}(\Phi^a) \quad \forall q.$$

In particular, $C^0(\Phi, \infty) = 0$.

When studying the existence and multiplicity of critical points of a functional we may often assume without loss of generality that there are only finitely many critical points. The following proposition relating the critical groups of Φ at infinity to those of its (finite) critical points is then immediate from Proposition 3.13 with $b = +\infty$.

Proposition 3.16. *If Φ satisfies* (C)*, then*

$$\dim C^q(\Phi, \infty) \leqslant \sum_{u \in K} \dim C^q(\Phi, u) \quad \forall q.$$

In particular, Φ has a critical point u with $C^q(\Phi, u) \neq 0$ when $C^q(\Phi, \infty) \neq 0$.

3.5. Minimizers and Maximizers

In this section we give sufficient conditions for Φ to have a global extremum and compute the critical groups at an isolated local extremum.

Proposition 3.17. *If Φ is bounded from below (resp. above) and satisfies* $(\mathrm{C})_c$ *for* $c = \inf \Phi$ (resp. $\sup \Phi$)*, then Φ has a global minimizer (resp. maximizer).*

PROOF. Apply Proposition 3.10 with $\mathcal{F} = \big\{ \{u\} : u \in W \big\}$ (resp. $\{W\}$). □

Turning to critical groups, we have

Proposition 3.18. *If u is an isolated local minimizer (resp. maximizer) of Φ, then*

$$C^q(\Phi, u) \approx \delta_{q0}\, \mathbb{Z}_2 \text{ (resp. } \delta_{qd}\, \mathbb{Z}_2).$$

PROOF. Let $c = \Phi(u)$. If u is an isolated local minimizer, then for sufficiently small $r > 0$,

$$\Phi(v) \geqslant c \quad \forall v \in B_r(u)$$

and there are no other critical points of Φ in $B_r(u)$. Since any $v \in B_r(u)$ with $\Phi(v) = c$ is then a local minimizer,

$$\Phi(v) > c, \quad v \in B_r(u) \backslash \{u\},$$

and hence

$$C^q(\Phi, u) = H^q(\Phi^c \cap B_r(u), \Phi^c \cap B_r(u) \backslash \{u\}) = H^q(\{u\}, \varnothing) \approx \delta_{q0}\, \mathbb{Z}_2$$

by (c_4).

If u is an isolated local maximizer, then for small $r > 0$,

$$\Phi(v) \leqslant c \quad \forall v \in B_r(u),$$

and hence

$$C^q(\Phi, u) = H^q(B_r(u), B_r(u) \backslash \{u\}) \approx \delta_{qd}\, \mathbb{Z}_2$$

by Example 2.5. □

Combining Propositions 3.15 (i), 3.17, and 3.18 gives

Corollary 3.19. *If Φ is bounded from below, satisfies* (C), *and has only a finite number of critical points, then $C^q(\Phi, \infty) \approx \delta_{q0}\, \mathbb{Z}_2$ and Φ has a global minimizer u with $C^q(\Phi, u) \approx \delta_{q0}\, \mathbb{Z}_2$.*

3.6. Homotopical Linking

The notion of homotopical linking is useful for obtaining critical points via the minimax principle.

Definition 3.20. Let A be a closed proper subset of a topological space X, $g \in C(A, W)$ such that $g(A)$ is closed and bounded, B a nonempty closed subset of W such that $\operatorname{dist}(g(A), B) > 0$, and

$$\Gamma = \{\gamma \in C(X, W) : \gamma(X) \text{ is closed, } \gamma|_A = g\}.$$

We say that (A, g) homotopically links B with respect to X if

$$\gamma(X) \cap B \neq \varnothing \quad \forall \gamma \in \Gamma. \tag{3.26}$$

When $g : A \subset W$ is the inclusion and $X = IA$, we will simply say that A homotopically links B.

Proposition 3.21. *If (A, g) homotopically links B with respect to X,*

$$c := \inf_{\gamma \in \Gamma} \sup_{u \in \gamma(X)} \Phi(u) \tag{3.27}$$

is finite, $\sup \Phi(g(A)) \leqslant \inf \Phi(B)$, *and Φ satisfies* $(\mathrm{C})_c$, *then* $c \geqslant \inf \Phi(B)$ *and is a critical value of Φ. If* $c = \inf \Phi(B)$, *then Φ has a critical point with critical value c on B.*

PROOF. By (3.26), $c \geqslant \inf \Phi(B)$. First suppose that $c > \inf \Phi(B)$, and let $2\varepsilon < c - \sup \Phi(g(A))$. Then for any $\eta \in \mathcal{D}_{c,\varepsilon}$ and all $t \in [0,1]$, $\eta(t,\cdot)$ is the identity on $g(A)$ by (iii) in the definition of $\mathcal{D}_{c,\varepsilon}$, so Γ is invariant under $\mathcal{D}_{c,\varepsilon}$ and hence Proposition 3.11 applies.

Now suppose that $c = \inf \Phi(B)$ and $K^c \cap B = \varnothing$. Since K^c is compact by $(\mathrm{C})_c$ and B is closed, $\operatorname{dist}(K^c, B) > 0$. Applying Lemma 3.7 to $-\Phi$ with $C = g(A) \cup K^c$ and $\delta < \operatorname{dist}(C, B)$ gives an $\varepsilon > 0$ and a homeomorphism η of W such that η is the identity outside $\Phi_{c-2\varepsilon}^{c+2\varepsilon} \backslash N_{\delta/3}(C)$ and

$$\eta(\Phi_{c-\varepsilon} \backslash N_\delta(C)) \subset \Phi_{c+\varepsilon},$$

in particular, η is the identity on $g(A)$ and $\Phi \geqslant c + \varepsilon$ on $\eta(B)$. Then taking a $\gamma \in \Gamma$ with $\Phi < c + \varepsilon$ on $\gamma(X)$, we have $\widetilde{\gamma} := \eta^{-1} \circ \gamma \in \Gamma$ and hence

$$\gamma(X) \cap \eta(B) = \eta(\widetilde{\gamma}(X) \cap B) \neq \varnothing$$

by (3.26), a contradiction. $\square$

To construct examples of homotopically linking sets, let $\mathcal{M}$ be a bounded symmetric subset of $W \backslash \{0\}$ radially homeomorphic to the unit sphere S, i.e., the restriction of the radial projection π_S to $\mathcal{M}$ is a homeomorphism. Then the radial projection from $W \backslash \{0\}$ onto $\mathcal{M}$ is given by

$$\pi_{\mathcal{M}} = \left(\pi_S|_{\mathcal{M}}\right)^{-1} \circ \pi_S.$$

Let $A_0 \neq \varnothing$ and B_0 be disjoint closed symmetric subsets of $\mathcal{M}$ such that

$$i(A_0) = i(\mathcal{M} \backslash B_0) < \infty \tag{3.28}$$

where i denotes the cohomological index.

Proposition 3.22. *Then $A = RA_0$ homotopically links $B = \pi_{\mathcal{M}}^{-1}(B_0) \cup \{0\}$ for any $R > 0$.*

PROOF. If not, there is a $\gamma \in C(IA, B^c)$ with $\gamma|_A = id_A$. Then

$$SA_0 \to \mathcal{M} \backslash B_0, \quad (u,t) \mapsto \begin{cases} \pi_{\mathcal{M}}(\gamma((1-t)\,Ru)), & t \in [0,1] \\ -\pi_{\mathcal{M}}(\gamma(-(1+t)\,Ru)), & t \in [-1,0) \end{cases}$$

is an odd map and hence

$$i(\mathcal{M} \backslash B_0) \geqslant i(SA_0) = i(A_0) + 1$$

by (i_2) and (i_6), contradicting (3.28). $\square$

Proposition 3.23. *If $h \in C(CA_0, \mathcal{M})$ is such that $h(CA_0)$ is closed in $\mathcal{M}$ and $h|_{A_0} = id_{A_0}$, then $A = R(IA_0 \cup h(CA_0))$ homotopically links $B = rB_0$ for any $0 < r < R$.*

PROOF. If not, there is a $\gamma \in C(IA, B^c)$ such that $\gamma(IA)$ is closed and $\gamma|_A = id_A$. Since $h|_{A_0} = id_{A_0}$ and A_0 is symmetric, h extends to an odd map $\widetilde{h} \in C(SA_0, \mathcal{M})$, and since $RIA_0 \subset A$ is symmetric, γ extends to an odd map $\widetilde{\gamma} \in C(I\widetilde{A}, B^c)$ where $\widetilde{A} = R\widetilde{h}(SA_0)$. Since $\widetilde{\gamma}(I\widetilde{A}) \cap r\mathcal{M} \subset r\mathcal{M} \backslash B$,

$$i(\widetilde{\gamma}(I\widetilde{A}) \cap r\mathcal{M}) \leqslant i(r\mathcal{M} \backslash B) = i(\mathcal{M} \backslash B_0) \tag{3.29}$$

by (i_2). Noting that $\widetilde{\gamma}|_{\widetilde{A}} = id_{\widetilde{A}}$ and applying Corollary 2.13 with $U = rI\mathcal{M}$ and $\psi = \widetilde{\gamma}$ gives

$$i(\widetilde{\gamma}(I\widetilde{A}) \cap r\mathcal{M}) \geqslant i(\widetilde{A}) \geqslant i(SA_0) = i(A_0) + 1 \tag{3.30}$$

by (i_2) and (i_6). Together, (3.29) and (3.30) contradict (3.28). □

3.7. Cohomological Linking

The notion of cohomological linking is useful for obtaining pairs of sublevel sets with nontrivial cohomology and hence critical points with nontrivial critical groups via Proposition 3.13.

Definition 3.24. Let A and B be disjoint nonempty subsets of W. We say that A cohomologically links B in dimension $q < \infty$ if

$$i^* : \widetilde{H}^q(B^c) \to \widetilde{H}^q(A),$$

induced by $i : A \subset B^c$, is nontrivial.

Proposition 3.25. *If A cohomologically links B in dimension q and*

$$\Phi|_A \leqslant a < \Phi|_B, \tag{3.31}$$

then $H^0(W, \Phi^a) = 0$ and $H^{q+1}(W, \Phi^a) \neq 0$. If, in addition, a is less than all critical values and Φ has only a finite number of critical points in $\Phi^{-1}(a, \infty)$ and satisfies (C)*, then $C^0(\Phi, \infty) = 0$, $C^{q+1}(\Phi, \infty) \neq 0$, and Φ has a critical point u with $\Phi(u) > a$ and $C^{q+1}(\Phi, u) \neq 0$.*

Proof. We have the commutative diagram

$$\begin{array}{ccc} \widetilde{H}^q(B^c) & \longrightarrow & \widetilde{H}^q(\Phi^a) \\ & \overset{i^*}{\searrow} & \downarrow \\ & & \widetilde{H}^q(A) \end{array}$$

induced by $A \subset \Phi^a \subset B^c$. Since $i^* \neq 0$, $\widetilde{H}^q(\Phi^a) \neq 0$ and hence the conclusions follow from Propositions 2.4 (ii) and 3.16. □

To construct an example, let $\mathcal{M}$, $\pi_{\mathcal{M}}$, A_0, and B_0 be as in the last section.

Proposition 3.26. *Then $A = RA_0$ cohomologically links $B = \pi_{\mathcal{M}}^{-1}(B_0) \cup \{0\}$ in dimension $q = i(A_0) - 1$ for any $R > 0$.*

Proof. We have the commutative diagram

$$\begin{array}{ccc} \widetilde{H}^q(\mathcal{M} \backslash B_0) & \xrightarrow{j^*} & \widetilde{H}^q(A_0) \\ \approx \downarrow & & \approx \downarrow \\ \widetilde{H}^q(B^c) & \xrightarrow{i^*} & \widetilde{H}^q(A) \end{array}$$

where $j : A_0 \subset \mathcal{M}\backslash B_0$ and the vertical maps are isomorphisms induced by the homotopy equivalence $\pi_{\mathcal{M}}|_{B^c} : B^c \to \mathcal{M}\backslash B_0$ and the homeomorphism $\pi_{\mathcal{M}}|_A : A \to A_0$. The conclusion follows since $j^* \neq 0$ by Proposition 2.14 *(iv)*. □

If A cohomologically links B in dimension q and h is a homeomorphism of W, then the following commutative diagram, where $j : h(A) \subset h(B^c)$, shows that $h(A)$ cohomologically links $h(B)$ in dimension q as well.

$$\begin{array}{ccc} \widetilde{H}^q(h(B^c)) & \xrightarrow{j^*} & \widetilde{H}^q(h(A)) \\ {\scriptstyle h^*}\downarrow{\scriptstyle \approx} & & {\scriptstyle h^*}\downarrow{\scriptstyle \approx} \\ \widetilde{H}^q(B^c) & \xrightarrow{i^*} & \widetilde{H}^q(A) \end{array}$$

In particular, we have the following generalization of Proposition 3.26.

Proposition 3.27. *If h is a homeomorphism of W, then $A = h(RA_0)$ cohomologically links $B = h(\pi_{\mathcal{M}}^{-1}(B_0) \cup \{0\})$ in dimension $q = i(A_0) - 1$ for any $R > 0$.*

3.8. Nontrivial Critical Points

In many applications Φ has the trivial critical point $u = 0$ and we are interested in finding others. We assume that Φ has only a finite number of critical points. The following proposition is useful for obtaining a nontrivial critical point with a nontrivial critical group.

Proposition 3.28. *Assume that Φ satisfies* (C).

(*i*) *If $C^q(\Phi, 0) = 0$ and $C^q(\Phi, \infty) \neq 0$ for some q, then Φ has a critical point $u \neq 0$ with $C^q(\Phi, u) \neq 0$.*

(*ii*) *If $C^q(\Phi, 0) \neq 0$ and $C^q(\Phi, \infty) = 0$ for some q, then Φ has a critical point $u \neq 0$ with either $\Phi(u) < 0$ and $C^{q-1}(\Phi, u) \neq 0$, or $\Phi(u) > 0$ and $C^{q+1}(\Phi, u) \neq 0$.*

First a purely topological lemma.

Lemma 3.29. *If $X_1 \subset X_2 \subset X_3 \subset X_4$ are topological spaces, then*

$$\dim H^{q-1}(X_2, X_1) + \dim H^{q+1}(X_4, X_3) \geqslant \dim H^q(X_3, X_2) - \dim H^q(X_4, X_1) \quad \forall q.$$

Proof. From the exact sequence

$$\cdots \longrightarrow H^{q-1}(X_2, X_1) \xrightarrow{\delta} H^q(X_3, X_2) \xrightarrow{j^*} H^q(X_3, X_1) \longrightarrow \cdots$$

of the triple (X_3, X_2, X_1), we have

$$\dim H^{q-1}(X_2, X_1) \geqslant \operatorname{rank} \delta = \operatorname{nullity} j^* = \dim H^q(X_3, X_2) - \operatorname{rank} j^* \geqslant \dim H^q(X_3, X_2) - \dim H^q(X_3, X_1),$$

and from the exact sequence

$$\cdots \longrightarrow H^q(X_4, X_1) \xrightarrow{i^*} H^q(X_3, X_1) \xrightarrow{\delta} H^{q+1}(X_4, X_3) \longrightarrow \cdots$$

of the triple (X_4, X_3, X_1),

$$\dim H^q(X_3, X_1) = \operatorname{rank}\delta + \operatorname{nullity}\delta \leqslant \dim H^{q+1}(X_4, X_3) + \operatorname{rank} i^* \leqslant \dim H^{q+1}(X_4, X_3) + \dim H^q(X_4, X_1),$$

so the conclusion follows. □

PROOF OF PROPOSITION 3.28. (*i*) By Proposition 3.16, Φ has a critical point u with $C^q(\Phi, u) \neq 0$ since $C^q(\Phi, \infty) \neq 0$, and $u \neq 0$ since $C^q(\Phi, 0) = 0$.

(*ii*) Let $\varepsilon > 0$ be so small that zero is the only critical value in $[-\varepsilon, \varepsilon]$ and a be less than $-\varepsilon$ and all critical values. Since $\dim H^q(\Phi^\varepsilon, \Phi^{-\varepsilon}) \geqslant \dim C^q(\Phi, 0)$ by Proposition 3.12 and $H^q(W, \Phi^a) = C^q(\Phi, \infty)$, applying Lemma 3.29 to $\Phi^a \subset \Phi^{-\varepsilon} \subset \Phi^\varepsilon \subset W$ gives

$$\dim H^{q-1}(\Phi^{-\varepsilon}, \Phi^a) + \dim H^{q+1}(W, \Phi^\varepsilon) \geqslant \dim C^q(\Phi, 0) - \dim C^q(\Phi, \infty) > 0.$$

Then either $H^{q-1}(\Phi^{-\varepsilon}, \Phi^a) \neq 0$, or $H^{q+1}(W, \Phi^\varepsilon) \neq 0$, and the conclusion follows from Proposition 3.13. □

Remark 3.30. The alternative in Proposition 3.28 (*ii*) and Lemma 3.29 were proved by Perera [**95, 96**].

3.9. Mountain Pass Points

A critical point u of Φ with $C^1(\Phi, u) \neq 0$ is called a mountain pass point. Since cohomology groups, and hence also critical groups, are trivial in negative dimensions, the special case $q = 0$ of Proposition 3.28 (*ii*) reduces to

Corollary 3.31. *If $C^0(\Phi, 0) \neq 0$, $C^0(\Phi, \infty) = 0$, and Φ satisfies* (C), *then Φ has a mountain pass point $u \neq 0$ with $\Phi(u) > 0$.*

This implies the well-known mountain pass lemma of Ambrosetti and Rabinowitz [**7**]. Indeed, if the origin is a local minimizer and Φ is unbounded from below, then $C^0(\Phi, 0) \approx \mathbb{Z}_2$ by Proposition 3.18 and $C^0(\Phi, \infty) = 0$ by Proposition 3.15 (*ii*), so Corollary 3.31 gives a positive mountain pass level.

3.10. Three Critical Points Theorem

Another consequence of Proposition 3.28 (*ii*) is

Corollary 3.32. *If $C^q(\Phi, 0) \neq 0$ for some $q \geqslant 1$ and Φ is bounded from below and satisfies* (C), *then Φ has a critical point $u_1 \neq 0$. If $q \geqslant 2$, then there is a second critical point $u_2 \neq 0$.*

PROOF. By Corollary 3.19, $C^q(\Phi, \infty) = 0$ and Φ has a global minimizer u_1 with $C^q(\Phi, u_1) = 0$. Since $C^q(\Phi, 0) \neq 0$, $u_1 \neq 0$ and there is a critical point $u_2 \neq 0$ with either $C^{q-1}(\Phi, u_2) \neq 0$ or $C^{q+1}(\Phi, u_2) \neq 0$. When $q \geqslant 2$, $u_2 \neq u_1$ since $C^{q-1}(\Phi, u_1) = C^{q+1}(\Phi, u_1) = 0$. $\square$

3.11. Cohomological Local Splitting

The notion of cohomological local splitting is useful for obtaining nontrivial critical groups at zero and hence nontrivial critical points via Proposition 3.28 *(ii)*.

Definition 3.33. We say that Φ has a cohomological local splitting near zero in dimension q, $1 \leqslant q < \infty$ if there are

(i) a bounded symmetric subset $\mathcal{M}$ of $W \setminus \{0\}$ that is radially homeomorphic to the unit sphere, and disjoint symmetric subsets $A_0 \neq \emptyset$ and B_0 of $\mathcal{M}$ such that

$$i(A_0) = i(\mathcal{M} \setminus B_0) = q, \tag{3.32}$$

(ii) a homeomorphism h from $I\mathcal{M}$ onto a neighborhood U of zero containing no other critical points, such that $h(0) = 0$ and

$$\Phi|_A \leqslant 0 < \Phi|_{B \setminus \{0\}} \tag{3.33}$$

where $A = h(IA_0)$ and $B = h(IB_0) \cup \{0\}$.

Proposition 3.34. *If Φ has a cohomological local splitting near zero in dimension q, then $C^q(\Phi, 0) \neq 0$.*

PROOF. (3.33) gives the commutative diagram

$$\begin{array}{ccccccc}
\Phi^0 \cap U & \xleftarrow{\;i\;} & \Phi^0 \cap U \setminus \{0\} & \longrightarrow & U \setminus B & \xleftarrow{\;h\;} & I\mathcal{M} \setminus (IB_0 \cup \{0\}) \\
\uparrow & & {\scriptstyle l}\uparrow & & {\scriptstyle \widetilde{k}}\uparrow & & {\scriptstyle k}\uparrow \\
A & \longleftarrow & h(A_0) & \xrightarrow{\;\widetilde{j}\;} & h(\mathcal{M} \setminus B_0) & \xleftarrow{\;h\;} & \mathcal{M} \setminus B_0 \\
{\scriptstyle h}\uparrow & & {\scriptstyle h}\uparrow & & {\scriptstyle h}\uparrow & & \\
IA_0 & \longleftarrow & A_0 & \xrightarrow{\;j\;} & \mathcal{M} \setminus B_0 & &
\end{array}$$

where i, j, $\widetilde{j}$, k, $\widetilde{k}$, l, and the unlabeled maps are inclusions. Passing to the induced diagram on reduced cohomology in dimension $q - 1$, $\widetilde{j}^* \neq 0$ since $j^* \neq 0$ by (3.32) and Proposition 2.14 *(iv)* and the h^* are isomorphisms. Since the radial projection onto $\mathcal{M}$ is a homotopy equivalence, k^*, and hence also $\widetilde{k}^*$, is an isomorphism. So the top middle square gives $l^* \neq 0$. On the other hand, the top left square gives $l^* i^* = 0$ since IA_0, and hence also A, is contractible. So i^* is not onto. Therefore δ, and hence also $C^q(\Phi, 0)$, is nontrivial in the exact sequence

$$\cdots \longrightarrow \widetilde{H}^{q-1}(\Phi^0 \cap U) \xrightarrow{i^*} \widetilde{H}^{q-1}(\Phi^0 \cap U \setminus \{0\}) \xrightarrow{\delta} H^q(\Phi^0 \cap U, \Phi^0 \cap U \setminus \{0\}) \longrightarrow \cdots$$

of the pair $(\Phi^0 \cap U, \Phi^0 \cap U \setminus \{0\})$. □

Remark 3.35. Definition 3.33 is a variant of the notion of homological local linking introduced by Perera [**96**], which also yields a nontrivial critical group at zero.

3.12. Even Functionals and Multiplicity

In this section we assume that Φ is even and use the cohomological index to obtain multiple critical points.

Let $\mathcal{F}$ denote the class of symmetric subsets of $W \setminus \{0\}$, and for $k \leqslant d$ in $\mathbb{N}$, let

$$\mathcal{F}_k = \{M \in \mathcal{F} : i(M) \geqslant k\} \tag{3.34}$$

and

$$c_k = \inf_{M \in \mathcal{F}_k} \sup_{u \in M} \Phi(u). \tag{3.35}$$

Since $\mathcal{F}_k \supset \mathcal{F}_{k+1}$, $c_k \leqslant c_{k+1}$, and since $S_\delta(0) \in \mathcal{F}_k$ for any $\delta > 0$ by (i_8) and $\sup \Phi(S_\delta(0)) \to 0$ as $\delta \searrow 0$ by continuity, $c_k \leqslant 0$.

Proposition 3.36. *Assume that Φ is even.*

(i) If $-\infty < c_k = \cdots = c_{k+m-1} = c < 0$ and Φ satisfies $(\mathrm{C})_c$, then $i(K^c) \geqslant m$. In particular, if $-\infty < c_k \leqslant \cdots \leqslant c_{k+m-1} < 0$ and Φ satisfies $(\mathrm{C})_c$ for $c = c_k, \dots, c_{k+m-1}$, then each c is a critical value and Φ has m distinct pairs of associated critical points.

(ii) If $-\infty < c_k < 0$ for all sufficiently large k and Φ satisfies $(\mathrm{C})_c$ for all $c < 0$, then $c_k \nearrow 0$.

First a lemma.

Lemma 3.37. *If $c < 0$ and Φ is even and satisfies $(\mathrm{C})_c$, then there is an $\varepsilon > 0$ such that*

$$c - \varepsilon < c_k \leqslant \cdots \leqslant c_{k+m-1} < c + \varepsilon \implies i(K^c) \geqslant m. \tag{3.36}$$

Proof. Since K^c is compact by $(\mathrm{C})_c$, there is a $\delta > 0$ such that

$$i(N_\delta(K^c)) = i(K^c) \tag{3.37}$$

by (i_4), and there are an $\varepsilon > 0$ and an odd map $\eta \in C(W, W)$ such that

$$\eta(\Phi^{c+\varepsilon} \setminus N_\delta(K^c)) \subset \Phi^{c-\varepsilon} \tag{3.38}$$

by Lemma 3.7. Then

$$i(\Phi^{c+\varepsilon}) \leqslant i(\overline{\Phi^{c+\varepsilon} \setminus N_\delta(K^c)}) + i(N_\delta(K^c)) \leqslant i(\Phi^{c-\varepsilon}) + i(K^c) \tag{3.39}$$

by (i_2), (i_5), and (3.37). If $c - \varepsilon < c_k$, then $\Phi^{c-\varepsilon} \notin \mathcal{F}_k$ and hence

$$i(\Phi^{c-\varepsilon}) \leqslant k - 1,$$

and if $c + \varepsilon > c_{k+m-1}$, then there is an $M \in \mathcal{F}_{k+m-1}$ with $M \subset \Phi^{c+\varepsilon}$ and hence

$$i(\Phi^{c+\varepsilon}) \geqslant i(M) \geqslant k + m - 1$$

by (i_2), so (3.36) follows from (3.39). □

PROOF OF PROPOSITION 3.36. (i) Taking $c = c_k = \cdots = c_{k+m-1}$ in Lemma 3.37 gives $i(K^c) \geqslant m$. Then taking $m = 1$ gives $i(K^{c_k}) \geqslant 1$ and hence $K^{c_k} \neq \varnothing$ by (i_1), so each c_k is a critical value. Either they are all distinct and therefore have distinct pairs of critical points, or some c is repeated, which then has $i(K^c) \geqslant 2$ and therefore infinitely many pairs of associated critical points by Proposition 2.14 (i).

(ii) If $c_k \nearrow c < 0$, then taking $\varepsilon > 0$ as in Lemma 3.37 and k so large that $c_k > c - \varepsilon$ gives $i(K^c) = \infty$ since $c_{k+m-1} \leqslant c$ for all m, contradicting the compactness of K^c by Proposition 2.14 (ii). □

Proposition 3.38. *If $A, B^c \in \mathcal{F}$ with*

$$i(A) \geqslant k + m - 1, \qquad i(B^c) \leqslant k - 1$$

for some $k, m \geqslant 1$ and

$$-\infty < a := \inf \Phi(B) \leqslant \sup \Phi(A) =: b < 0,$$

and Φ is even and satisfies $(\mathrm{C})_c$ for all $c \in [a, b]$, then $a \leqslant c_k \leqslant \cdots \leqslant c_{k+m-1} \leqslant b$ and hence there are m distinct pairs of critical points in Φ_a^b.

PROOF. Each $M \in \mathcal{F}_k$ satisfies $i(M) > i(B^c)$ and hence intersects B by (i_2), and $\Phi \geqslant a$ on B, so $c_k \geqslant a$. Since $A \in \mathcal{F}_{k+m-1}$ and $\Phi \leqslant b$ on A, $c_{k+m-1} \leqslant b$. So the conclusion follows from Proposition 3.36. □

Remark 3.39. Since $A \in \mathcal{F}_{k+m-1} \subset \mathcal{F}_k$ intersects B, $\inf \Phi(B) \leqslant \sup \Phi(A)$.

3.13. Pseudo-Index

The notion of a pseudo-index introduced by Benci [**16**] is useful for, among other things, obtaining multiple critical points of an even functional at positive levels.

Definition 3.40. Let $\mathcal{F}$ be as in the last section, $\mathcal{M} \in \mathcal{F}$ be closed, $0 \leqslant a < b \leqslant +\infty$, and denote by Γ the group of odd homeomorphisms of W that are the identity outside $\Phi^{-1}(a, b)$. Then the pseudo-index of $M \in \mathcal{F}$ related to i, $\mathcal{M}$, and Γ is defined by

$$i^*(M) = \min_{\gamma \in \Gamma} i(\gamma(M) \cap \mathcal{M}).$$

The following proposition lists some properties of the pseudo-index $i^* : \mathcal{F} \to \mathbb{N} \cup \{0, \infty\}$.

Proposition 3.41. *Let $A, B \in \mathcal{F}$.*

(*i*) *If* $A \subset B$, *then*

$$i^*(A) \leqslant i^*(B).$$

(*ii*) *If* $\eta \in \Gamma$, *then*

$$i^*(\eta(A)) = i^*(A).$$

(*iii*) *If* A *and* B *are closed, then*

$$i^*(A \cup B) \leqslant i^*(A) + i(B).$$

PROOF. (*i*) For each $\gamma \in \Gamma$,

$$i(\gamma(A) \cap \mathcal{M}) \leqslant i(\gamma(B) \cap \mathcal{M})$$

by (i_2).

(*ii*) We have

$$i^*(\eta(A)) = \min_{\gamma \in \Gamma} i(\gamma\eta(A) \cap \mathcal{M}) = i^*(A)$$

since $\{\gamma\eta : \gamma \in \Gamma\} = \Gamma$.

(*iii*) For each $\gamma \in \Gamma$,

$$\begin{aligned} i(\gamma(A \cup B) \cap \mathcal{M}) &= i((\gamma(A) \cap \mathcal{M}) \cup (\gamma(B) \cap \mathcal{M})) \\ &\leqslant i(\gamma(A) \cap \mathcal{M}) + i(\gamma(B) \cap \mathcal{M}) \end{aligned}$$

by (i_5) and

$$i(\gamma(B) \cap \mathcal{M}) \leqslant i(\gamma(B)) = i(B)$$

by (i_2). □

For $k \leqslant i(\mathcal{M})$ in $\mathbb{N}$, let

$$\mathcal{F}_k^* = \{M \in \mathcal{F} : i^*(M) \geqslant k\}$$

and

$$c_k^* = \inf_{M \in \mathcal{F}_k^*} \sup_{u \in M} \Phi(u).$$

Since $\mathcal{F}_k^* \supset \mathcal{F}_{k+1}^*$, $c_k^* \leqslant c_{k+1}^*$.

Proposition 3.42. *Assume that* Φ *is even.*

(*i*) *If* $a < c_k^* = \cdots = c_{k+m-1}^* = c < b$ *and* Φ *satisfies* $(\mathrm{C})_c$, *then* $i(K^c) \geqslant m$. *In particular, if* $a < c_k^* \leqslant \cdots \leqslant c_{k+m-1}^* < b$ *and* Φ *satisfies* $(\mathrm{C})_c$ *for* $c = c_k^*, \dots, c_{k+m-1}^*$, *then each* c *is a critical value and* Φ *has* m *distinct pairs of associated critical points.*

(*ii*) *If* $a = 0$, $b = +\infty$, $0 < c_k^* < +\infty$ *for all sufficiently large* k, *and* Φ *satisfies* $(\mathrm{C})_c$ *for all* $c > 0$, *then* $c_k^* \nearrow +\infty$.

As in the proof of Proposition 3.36, it suffices to prove the following lemma.

Lemma 3.43. *If* $c > 0$ *and* Φ *is even and satisfies* $(\mathrm{C})_c$, *then there is an* $\varepsilon > 0$ *such that*

$$c - \varepsilon < c_k^* \leqslant \cdots \leqslant c_{k+m-1}^* < c + \varepsilon \implies i(K^c) \geqslant m.$$

PROOF. As in the proof of Lemma 3.37, there are $\delta, \varepsilon > 0$ and $\eta \in \Gamma$ such that (3.37) and (3.38) hold. So

$$i^*(\Phi^{c+\varepsilon}) \leqslant i^*(\overline{\Phi^{c+\varepsilon}\backslash N_\delta(K^c)}) + i(N_\delta(K^c)) \leqslant i^*(\Phi^{c-\varepsilon}) + i(K^c)$$

by Proposition 3.41, and the conclusion follows as before. □

Proposition 3.44. *If $\mathcal{M} \in \mathcal{F}$ is bounded and radially homeomorphic to the unit sphere, $U = I\mathcal{M}$, $A, B \in \mathcal{F}$ with $A \subset U^c$ compact and $B \subset \mathcal{M}$,*

$$i(A) \geqslant k + m - 1, \qquad i(\mathcal{M}\backslash B) \leqslant k - 1$$

for some $k, m \geqslant 1$, and

$$\sup \Phi(A) \leqslant a < \inf \Phi(B) \leqslant \sup \Phi(IA) < b,$$

and Φ is even and satisfies $(\mathrm{C})_c$ for all $c \in (a, b)$, then $a < c_k^* \leqslant \cdots \leqslant c_{k+m-1}^* < b$ and hence there are m distinct pairs of critical points in $\Phi^{-1}(a, b)$.

PROOF. We apply Proposition 3.42.

Each $M \in \mathcal{F}_k^*$ satisfies

$$i(M \cap \mathcal{M}) \geqslant i^*(M) > i(\mathcal{M}\backslash B)$$

and hence intersects B by (i_2), and $\inf \Phi(B) > a$, so $c_k^* > a$.

If $\gamma \in \Gamma$, noting that $\gamma|_A = id_A$ since $\Phi \leqslant a$ on A and applying Corollary 2.13 with $\psi = \gamma|_{IA}$ gives

$$i(\gamma(IA) \cap \mathcal{M}) \geqslant i(A) \geqslant k + m - 1.$$

So $i^*(IA) \geqslant k + m - 1$ and hence $IA \in \mathcal{F}_{k+m-1}^*$. Since $\sup \Phi(IA) < b$, then $c_{k+m-1}^* < b$. □

Remark 3.45. Since $IA \in \mathcal{F}_{k+m-1}^* \subset \mathcal{F}_k^*$ intersects B, $\inf \Phi(B) \leqslant \sup \Phi(IA)$.

3.14. Functionals on Finsler Manifolds

A C^1-manifold modeled on a Banach space V is a connected Hausdorff space $\mathcal{M}$ together with a collection $\{(U_\lambda, \varphi_\lambda)\}_{\lambda \in \Lambda}$, called an atlas on $\mathcal{M}$, of pairs $(U_\lambda, \varphi_\lambda)$, called charts, such that

- (i) $\{U_\lambda\}_{\lambda \in \Lambda}$ is an open covering of $\mathcal{M}$,
- (ii) for each $\lambda \in \Lambda$, $\varphi_\lambda : U_\lambda \to \varphi_\lambda(U_\lambda) \subset V$ is a homeomorphism,
- (iii) for each $\lambda, \mu \in \Lambda$, $\varphi_\mu \circ \varphi_\lambda^{-1} : \varphi_\lambda(U_\lambda \cap U_\mu) \to \varphi_\mu(U_\lambda \cap U_\mu)$ is a C^1-diffeomorphism.

We refer to Lang [**64**] for the basic constructions on $\mathcal{M}$ such as the tangent bundle $T\mathcal{M}$ and the cotangent bundle $T^*\mathcal{M}$.

A Finsler manifold is a regular C^1-Banach manifold $\mathcal{M}$ together with a continuous function $\|\cdot\| : T\mathcal{M} \to [0, \infty)$, called a Finsler structure on $T\mathcal{M}$, such that

- (i) for each $u \in \mathcal{M}$, the restriction $\|\cdot\|_u$ of $\|\cdot\|$ to the tangent space $T_u\mathcal{M}$ at u is a norm, (*ii*) for each $u \in \mathcal{M}$ and $C > 1$, there is a trivializing neighborhood U of u such that

$$\frac{1}{C} \left\| \cdot \right\|_v \leqslant \left\| \cdot \right\|_u \leqslant C \left\| \cdot \right\|_v \quad \forall v \in U.$$

Then $T^*\mathcal{M}$ has the dual Finsler structure

$$\left\| w \right\|_u^* = \sup_{\substack{v \in T_u\mathcal{M} \\ \left\| v \right\|_u = 1}} \langle w, v \rangle, \quad w \in T_u^*\mathcal{M} \tag{3.40}$$

where $\langle \cdot, \cdot \rangle$ is the duality pairing between the cotangent space $T_u^*\mathcal{M}$ and $T_u\mathcal{M}$.

The Finsler structure also induces a metric

$$d(u, v) = \inf_{\substack{\sigma \in C^1([0,1],\mathcal{M}) \\ \sigma(0)=u,\, \sigma(1)=v}} \int_0^1 \left\| \sigma'(t) \right\|_{\sigma(t)} dt$$

on $\mathcal{M}$, called the Finsler metric, that is consistent with the topology of $\mathcal{M}$. We assume that $\mathcal{M}$ is complete with respect to d.

If $\Phi \in C^1(\mathcal{M}, \mathbb{R})$, then $\Phi'(u) \in T_u^*\mathcal{M}$ and the terminology and notation at the beginning of this chapter and Definition 3.1 still apply. We refer to Corvellec, Degiovanni, and Marzocchi [**34**] for the proofs of

Lemma 3.46 (First Deformation Lemma). *If $c \in \mathbb{R}$, $\varepsilon_0, \delta > 0$, and Φ satisfies* (PS)$_c$, *then there are $\varepsilon > 0$ and $\eta \in C([0,1] \times \mathcal{M}, \mathcal{M})$ satisfying*

(*i*) $d(\eta(t,u), u) \leqslant \delta t \quad \forall (t,u) \in [0,1] \times \mathcal{M}$,
(*ii*) *$\eta(t, \cdot)$ is the identity outside $\Phi_{c-\varepsilon_0}^{c+\varepsilon_0}$ for all $t \in [0,1]$,*
(*iii*) $\Phi(\eta(t,u)) \leqslant \Phi(u) \quad \forall (t,u) \in [0,1] \times \mathcal{M}$,
(*iv*) $\eta(1, \Phi^{c+\varepsilon} \backslash N_\delta(K^c)) \subset \Phi^{c-\varepsilon}$.

Lemma 3.47 (Second Deformation Lemma). *If $-\infty < a < b \leqslant +\infty$ and Φ has no critical values in $[a,b]$ and satisfies* (PS)$_c$ *for all $c \in [a,b] \cap \mathbb{R}$, then Φ^a is a deformation retract of Φ^b.*

Much of the theory of the previous sections can now be adapted to this setting.

Proposition 3.48. *If $c := \inf \Phi(\mathcal{M})$ is finite and Φ satisfies* (PS)$_c$, *then c is a critical value of Φ.*

Proof. If not, let $\varepsilon > 0$ and $\eta \in C([0,1] \times \mathcal{M}, \mathcal{M})$ be given by Lemma 3.46 and take $u \in \mathcal{M}$ with $\Phi(u) \leqslant c + \varepsilon$. Then $\Phi(\eta(1,u)) \leqslant c - \varepsilon$ by (*iv*), a contradiction. □

Let A and B be disjoint nonempty subsets of $\mathcal{M}$ and

$$\Gamma = \left\{ \gamma \in C(CA, \mathcal{M}) : \gamma|_A = id_A \right\}.$$

We say that A homotopically links B if

$$\gamma(CA) \cap B \neq \varnothing \quad \forall \gamma \in \Gamma.$$

Proposition 3.49. *If $\mathcal{M}$ is a free $\mathbb{Z}_2$-space and A and B are disjoint nonempty symmetric subsets of $\mathcal{M}$ such that*

$$i(A) = i(\mathcal{M}\backslash B) < \infty, \tag{3.41}$$

then A homotopically links B.

PROOF. If not, there is a $\gamma \in C(CA, \mathcal{M}\backslash B)$ with $\gamma|_A = id_A$. Then

$$SA \to \mathcal{M}\backslash B, \quad (u,t) \mapsto \begin{cases} \gamma(u,t), & t \in [0,1] \\ -\gamma(-u,-t), & t \in [-1,0) \end{cases}$$

is an odd map and hence

$$i(\mathcal{M}\backslash B) \geqslant i(SA) = i(A) + 1$$

by (i_2) and (i_6), contradicting (3.41). □

Proposition 3.50. *If A homotopically links B,*

$$c := \inf_{\gamma \in \Gamma} \sup_{u \in \gamma(CA)} \Phi(u) \tag{3.42}$$

is finite, $\sup \Phi(A) < \inf \Phi(B)$, and Φ satisfies $(\mathrm{PS})_c$, then $c \geqslant \inf \Phi(B)$ is a critical value of Φ.

PROOF. If not, take $\varepsilon_0 < c - \sup \Phi(A)$, let ε and η be given by Lemma 3.46, and take $\gamma \in \Gamma$ with $\sup \Phi(\gamma(CA)) \leqslant c + \varepsilon$. Then $\eta(1,\cdot)$ is the identity on A by (ii), so $\eta(1,\cdot) \circ \gamma \in \Gamma$, but $\sup \Phi(\eta(1, \gamma(CA))) \leqslant c - \varepsilon$ by (iv), contradicting (3.42). □

Now suppose that $\mathcal{M}$ is a free $\mathbb{Z}_2$-space and Φ is even. The proof of the following symmetric deformation lemma can be found in Corvellec [**32**].

Lemma 3.51. *If $c \in \mathbb{R}$, $\delta > 0$, and Φ is even and satisfies $(\mathrm{PS})_c$, then there are $\varepsilon > 0$ and $\eta \in C([0,1] \times \mathcal{M}, \mathcal{M})$, with $\eta(t,\cdot)$ odd for all $t \in [0,1]$, satisfying (i), (iii), and (iv) of Lemma 3.46.*

Let $\mathcal{F}$ denote the class of invariant subsets of $\mathcal{M}$ and for $k \leqslant \dim \mathcal{M}$ in $\mathbb{N}$, let $\mathcal{F}_k$ and c_k be defined by (3.34) and (3.35), respectively. As in the proof of Lemma 3.37, it then follows from Lemma 3.51 that if $c \in \mathbb{R}$ and Φ satisfies $(\mathrm{PS})_c$, there is an $\varepsilon > 0$ such that

$$c - \varepsilon < c_k \leqslant \cdots \leqslant c_{k+m-1} < c + \varepsilon \implies i(K^c) \geqslant m.$$

This in turn gives

Proposition 3.52. *Assume that Φ is even.*

(i) If $-\infty < c_k = \cdots = c_{k+m-1} = c < +\infty$ and Φ satisfies $(\mathrm{PS})_c$, then $i(K^c) \geqslant m$. In particular, if $-\infty < c_k \leqslant \cdots \leqslant c_{k+m-1} < +\infty$ and Φ satisfies $(\mathrm{PS})_c$ for $c = c_k, \dots, c_{k+m-1}$, then each c is a critical value and Φ has m distinct pairs of associated critical points.

(ii) If $-\infty < c_k < +\infty$ for all sufficiently large k and Φ satisfies (PS), then $c_k \nearrow +\infty$.

The following proposition gives the indices of the sublevel sets of Φ.

Proposition 3.53. *Assume that Φ is even.*

(*i*) *If c_k is finite and Φ satisfies* $(\mathrm{PS})_{c_k}$*, then*

$$i(\mathcal{M}\backslash\Phi_{c_k}) < k \leqslant i(\Phi^{c_k}). \tag{3.43}$$

(*ii*) *If $c_k < c_{k+1}$ are finite and Φ satisfies* $(\mathrm{PS})_c$ *for $c = c_k, c_{k+1}$, then*

$$i(\Phi^{c_k}) = i(\mathcal{M}\backslash\Phi_a) = i(\Phi^a) = i(\mathcal{M}\backslash\Phi_{c_{k+1}}) = k \quad \forall a \in (c_k, c_{k+1}).$$

Proof. (*i*) By Proposition 2.14 (*iii*), $\mathcal{M}\backslash\Phi_{c_k}$ has a compact subset $C \in \mathcal{F}$ with $i(C) = i(\mathcal{M}\backslash\Phi_{c_k})$. Then $\max \Phi(C) < c_k$ since Φ is continuous, so $C \notin \mathcal{F}_k$ and hence $i(C) < k$, and the first inequality in (3.43) follows.

By (i_4), Φ^{c_k} has a closed neighborhood $N \in \mathcal{F}$ with $i(N) = i(\Phi^{c_k})$. Since K^{c_k} is a compact subset of Φ^{c_k},

$$\delta := \frac{1}{2} \operatorname{dist}(K^{c_k}, \mathcal{M}\backslash N) > 0.$$

By Lemma 3.51, there are an $\varepsilon > 0$ and an odd map $\eta \in C(\mathcal{M}, \mathcal{M})$ satisfying

$$d(\eta(u), u) \leqslant \delta \quad \forall u \in \mathcal{M}, \qquad \eta(\Phi^{c_k+\varepsilon}\backslash N_\delta(K^{c_k})) \subset \Phi^{c_k-\varepsilon}$$

and hence

$$\eta(\Phi^{c_k+\varepsilon}) \subset \Phi^{c_k-\varepsilon} \cup N_{2\delta}(K^{c_k}) \subset N.$$

Taking an $M \in \mathcal{F}_k$ with $M \subset \Phi^{c_k+\varepsilon}$ then gives

$$i(N) \geqslant i(\Phi^{c_k+\varepsilon}) \geqslant i(M) \geqslant k$$

by (i_2), so the second inequality in (3.43) also follows.

(*ii*) By (3.43) and (i_2),

$$k \leqslant i(\Phi^{c_k}) \leqslant i(\mathcal{M}\backslash\Phi_a) \leqslant i(\Phi^a) \leqslant i(\mathcal{M}\backslash\Phi_{c_{k+1}}) < k+1. \qquad \square$$

Often $\mathcal{M}$ is of the form

$$\mathcal{M} = \big\{u \in W : I(u) = 1\big\}$$

with 1 a regular value of $I \in C^1(W, \mathbb{R})$. Then $\mathcal{M}$ is a C^1-Finsler manifold by the implicit function theorem, and $\mathcal{M}$ is complete by the continuity of I. Moreover,

$$T_u\mathcal{M} = \big\{v \in W : \big\langle I'(u), v\big\rangle = 0\big\} = \ker I'(u).$$

Proposition 3.54. *If Ψ is a C^1-functional defined in a neighborhood of $\mathcal{M}$ and $\widetilde{\Psi}$ its restriction to $\mathcal{M}$, then the norm of $\widetilde{\Psi}'(u) \in T_u^*\mathcal{M}$ is given by*

$$\big\|\widetilde{\Psi}'(u)\big\|_u^* = \min_{\mu \in \mathbb{R}} \big\|\Psi'(u) - \mu\, I'(u)\big\|^*.$$

PROOF. We have

$$\begin{aligned}\big\|\widetilde{\Psi}'(u)\big\|_u^* &= \sup_{\substack{v\in\ker I'(u)\\ \|v\|=1}} \big\langle \Psi'(u), v\big\rangle && \text{by (3.40)}\\ &= \Big\| \,\Psi'(u)\big|_{\ker I'(u)} \Big\|^* \\ &= \min_{\mu\in\mathbb{R}} \big\|\Psi'(u) - \mu\, I'(u)\big\|^* && \text{by Lemma 3.55 below.} \quad \square\end{aligned}$$

Lemma 3.55. *If $L, M \in W^*$, then*

$$\big\| \,L|_{\ker M} \big\|^* = \min_{\mu\in\mathbb{R}} \|L - \mu M\|^* .$$

PROOF. For each $\mu \in \mathbb{R}$,

$$\big\| \,L|_{\ker M} \big\|^* = \sup_{\substack{v\in\ker M\\ \|v\|=1}} \langle L, v\rangle \leqslant \sup_{\|v\|=1} \langle L - \mu M, v\rangle = \|L - \mu M\|^* .$$

By the Hahn-Banach theorem, there is an $\widetilde{L} \in W^*$ such that $\widetilde{L} = L$ on $\ker M$ and

$$\big\|\widetilde{L}\big\|^* = \big\| \,L|_{\ker M} \big\|^*.$$

Since $\ker\big(L - \widetilde{L}\big) \supset \ker M$, $L - \widetilde{L} = \mu M$ for some $\mu \in \mathbb{R}$, so

$$\big\| \,L|_{\ker M} \big\|^* = \|L - \mu M\|^* . \quad \square$$

CHAPTER 4

p-Linear Eigenvalue Problems

In this chapter we study the p-linear eigenvalue problem

$$A_p u = \lambda B_p u \tag{4.1}$$

in W^*, where A_p satisfies (A_1) - (A_4), $B_p : W \to W^*$ is

(B_1) $(p-1)$-homogeneous and odd,

(B_2) strictly positive:

$$\langle B_p u, u \rangle > 0 \quad \forall u \neq 0,$$

(B_3) a compact potential operator,

and $\lambda \in \mathbb{R}$. We say that λ is an eigenvalue of (4.1) if there is a $u \neq 0$ in W satisfying (4.1), called an eigenvector associated with λ. Then αu is also an eigenvector associated with λ for any $\alpha \neq 0$ by (A_1) and (B_1), and

$$\lambda = \frac{\langle A_p u, u \rangle}{\langle B_p u, u \rangle} > 0 \tag{4.2}$$

by (A_2) and (B_2). The set $\sigma(A_p, B_p)$ of all eigenvalues is called the spectrum of the pair of operators (A_p, B_p).

Example 4.1. In problem (1.9), the usual choice is

$$\langle B_p u, v \rangle = \int_\Omega |u|^{p-2} uv :$$

$$\begin{cases} -\Delta_p u = \lambda |u|^{p-2} u & \text{in } \Omega \\ \\ u = 0 & \text{on } \partial\Omega. \end{cases}$$

Then the first eigenvalue λ_1 is positive, simple, and has an associated eigenfunction φ_1 that is positive in Ω (see Lindqvist [**68, 69**]). Moreover, λ_1 is isolated in the spectrum $\sigma(-\Delta_p)$, so the second eigenvalue $\lambda_2 = \inf \sigma(-\Delta_p) \cap (\lambda_1, \infty)$ is well-defined. In the ODE case $n = 1$, where Ω is an interval, the spectrum consists of a sequence of simple eigenvalues $\lambda_k \nearrow \infty$, and the eigenfunction φ_k associated with λ_k has exactly $k-1$ interior zeroes (see, e.g., Drábek [**46**]). In the semilinear PDE case $n \geqslant 2$, $p = 2$ also, $\sigma(-\Delta)$ consists of a sequence of eigenvalues $\lambda_k \nearrow \infty$. In the quasilinear PDE case $n \geqslant 2$, $p \neq 2$, increasing and unbounded sequences of eigenvalues can be defined using various minimax schemes, but a complete list of the eigenvalues of $-\Delta_p$ is still unavailable.

Our setting also includes weighted eigenvalue problems as the next example shows.

Example 4.2. In problem (1.9), taking

$$\big\langle B_p\, u, v\big\rangle = \int_\Omega V(x)\, |u|^{p-2}\, uv$$

with the weight $V(x) > 0$ a.e. and $V \in L^s(\Omega)$ for some

$$s \begin{cases} > \dfrac{n}{p}, & p \leqslant n \\ = 1, & p > n \end{cases}$$

gives

$$\begin{cases} -\Delta_p\, u = \lambda\, V(x)\, |u|^{p-2}\, u & \text{in } \Omega \\ \qquad u = 0 & \text{on } \partial\Omega. \end{cases}$$

This includes singular weights such as $V(x) = |x|^{-q}$, $0 < q < \min\{p, n\}$.

4.1. Variational Setting

Since

$$\big\langle I_p'(u), u\big\rangle = \langle A_p\, u, u\rangle = p\, I_p(u), \tag{4.3}$$

zero is the only critical value of I_p and hence it follows from the implicit function theorem that

$$\mathcal{M} := \big\{u \in W : I_p(u) = 1\big\} \tag{4.4}$$

is a C^1-Finsler manifold. Moreover, $\mathcal{M}$ is complete, symmetric, and radially homeomorphic to the unit sphere since I_p is continuous, even, and p-homogeneous. By (1.6),

$$\left(\frac{p}{C_0}\right)^{1/p} \leqslant \inf_{u \in \mathcal{M}} \|u\| \leqslant \sup_{u \in \mathcal{M}} \|u\| \leqslant \left(\frac{p}{c_0}\right)^{1/p}. \tag{4.5}$$

By Proposition 1.2, the potential J_p of B_p satisfying $J_p(0) = 0$ is given by

$$J_p(u) = \frac{1}{p}\, \big\langle B_p\, u, u\big\rangle$$

and is p-homogeneous and even. By (B_2), $J_p \geqslant 0$ and > 0 on $W \backslash \{0\}$. So the functional

$$\Psi(u) = \frac{1}{J_p(u)}, \quad u \in W \backslash \{0\}$$

is positive and its restriction

$$\widetilde{\Psi} = \Psi|_{\mathcal{M}} \tag{4.6}$$

is C^1. We will show that the critical values and the critical points of $\widetilde{\Psi}$ are the eigenvalues and the eigenvectors of (4.1).

Since

$$\Psi'(u) = -\frac{J_p'(u)}{J_p(u)^2} = -\widetilde{\Psi}(u)^2\, B_p\, u, \qquad I_p'(u) = A_p\, u,$$

it follows from Proposition 3.54 that the norm of $\widetilde{\Psi}'(u) \in T_u^*\mathcal{M}$ is given by

$$\left\|\widetilde{\Psi}'(u)\right\|_u^* = \min_{\mu \in \mathbb{R}} \left\|\mu\, A_p\, u + \widetilde{\Psi}(u)^2\, B_p\, u\right\|^*. \tag{4.7}$$

Lemma 4.3. *Eigenvalues of* (4.1) *coincide with the critical values of* $\widetilde{\Psi}$, *i.e.,* λ *is an eigenvalue if and only if there is a* $u \in \mathcal{M}$ *such that* $\widetilde{\Psi}'(u) = 0$ *and* $\widetilde{\Psi}(u) = \lambda$.

PROOF. By (4.7), $\widetilde{\Psi}'(u) = 0$ if and only if

$$\mu\, A_p\, u + \widetilde{\Psi}(u)^2\, B_p\, u = 0 \tag{4.8}$$

for some $\mu \in \mathbb{R}$. If (4.8) holds, applying it to u gives

$$\mu = -\widetilde{\Psi}(u)^2\, \frac{\langle B_p\, u, u\rangle}{\langle A_p\, u, u\rangle} = -\widetilde{\Psi}(u)^2\, \frac{J_p(u)}{I_p(u)} = -\widetilde{\Psi}(u) < 0,$$

so (4.8) reduces to (4.1) with $\lambda = \widetilde{\Psi}(u)$. Conversely, if λ is an eigenvalue and $u \in \mathcal{M}$ is an associated eigenvector,

$$\lambda = \frac{I_p(u)}{J_p(u)} = \widetilde{\Psi}(u)$$

by (4.2), so (4.1) implies (4.8) with $\mu = -\widetilde{\Psi}(u)$. □

Example 4.4. In Example 4.1,

$$\mathcal{M} = \left\{u \in W_0^{1,p}(\Omega) : \frac{1}{p}\int_\Omega |\nabla u|^p = 1\right\}, \qquad J_p(u) = \frac{1}{p}\int_\Omega |u|^p,$$

$$\widetilde{\Psi}(u) = \frac{p}{\displaystyle\int_\Omega |u|^p}.$$

Lemma 4.5. $\widetilde{\Psi}$ *satisfies* (PS).

PROOF. Let $(u_j) \subset \mathcal{M}$ be a $(\mathrm{PS})_c$ sequence, i.e.,

$$\widetilde{\Psi}(u_j) \to c, \quad \left\|\widetilde{\Psi}'(u_j)\right\|_{u_j}^* \to 0.$$

By (4.5), (u_j) is bounded, so a renamed subsequence converges weakly to some $u \in W$ since W is reflexive, and $B_p\, u_j$ converges to some $L \in W^*$ for a further subsequence since B_p is compact. Then

$$\widetilde{\Psi}(u_j) = \frac{p}{\langle B_p\, u_j, u_j\rangle} \to \frac{p}{\langle L, u\rangle} \neq 0$$

by Lemma 3.4, so $c \neq 0$.

By (4.7),

$$\mu_j\, A_p\, u_j + \widetilde{\Psi}(u_j)^2\, B_p\, u_j \to 0 \tag{4.9}$$

for some sequence $(\mu_j) \subset \mathbb{R}$. Applying (4.9) to u_j gives $\mu_j + \widetilde{\Psi}(u_j) \to 0$, so $\mu_j \to -c \neq 0$. Now applying (4.9) to $u_j - u$ gives $\langle A_p\, u_j, u_j - u\rangle \to 0$ since $\big\langle B_p\, u_j, u_j - u\big\rangle \to 0$ by Lemma 3.4, so $u_j \to u \in \mathcal{M}$ for a subsequence by (A_4). □

Lemma 4.5 implies that the set

$$E_\lambda = \big\{u \in \mathcal{M} : \widetilde{\Psi}'(u) = 0,\ \widetilde{\Psi}(u) = \lambda\big\}$$

of eigenvectors associated with λ that lie on $\mathcal{M}$ is compact and that the spectrum

$$\sigma(A_p\,, B_p\,) = \big\{\lambda \in \mathbb{R} : \widetilde{\Psi}'(u) = 0,\ \widetilde{\Psi}(u) = \lambda \text{ for some } u \in \mathcal{M}\big\}$$

is closed.

4.2. Minimax Eigenvalues

We now construct an unbounded sequence of minimax eigenvalues for the eigenvalue problem (4.1). Although this can be done using the Krasnoselskii genus as usual, we prefer to use the cohomological index in order to obtain a nontrivial critical group and later construct linking sets.

Let $\mathcal{F}$ denote the class of symmetric subsets of $\mathcal{M}$, let

$$\mathcal{K} = \big\{k \in \mathbb{N} : k \leqslant d\big\},$$

and for $k \in \mathcal{K}$, let

$$\mathcal{F}_k = \big\{M \in \mathcal{F} : i(M) \geqslant k\big\}$$

and

$$\lambda_k = \inf_{M \in \mathcal{F}_k}\ \sup_{u \in M}\ \widetilde{\Psi}(u).$$

Since $\mathcal{F}_k \supset \mathcal{F}_{k+1}$, $\lambda_k \leqslant \lambda_{k+1}$, and since the intersection of $\mathcal{M}$ with any k-dimensional subspace of W is a compact set in $\mathcal{F}_k$ by (i_8), λ_k is finite. When $d < \infty$ we set

$$\lambda_{d+1} = \infty$$

for convenience.

Theorem 4.6. *Assume (A_1) - (A_4) and (B_1) - (B_3). Then $(\lambda_k)_{k\in\mathcal{K}}$ is a nondecreasing sequence of eigenvalues of (4.1).*

(i) *If $\lambda_k = \cdots = \lambda_{k+m-1} = \lambda$, then $i(E_\lambda) \geqslant m$. In particular, there are d distinct pairs of eigenvectors on $\mathcal{M}$.*

(ii) *The smallest eigenvalue, called the first eigenvalue, is*

$$\lambda_1 = \min_{u \in \mathcal{M}} \widetilde{\Psi}(u) = \min_{u \neq 0} \frac{I_p(u)}{J_p(u)} > 0.$$

(*iii*) *We have* $i(\mathcal{M} \backslash \widetilde{\Psi}_{\lambda_k}) < k \leqslant i(\widetilde{\Psi}^{\lambda_k})$. *If* $\lambda_k < \lambda < \lambda_{k+1}$, *then*

$$i(\widetilde{\Psi}^{\lambda_k}) = i(\mathcal{M} \backslash \widetilde{\Psi}_\lambda) = i(\widetilde{\Psi}^\lambda) = i(\mathcal{M} \backslash \widetilde{\Psi}_{\lambda_{k+1}}) = k.$$

(*iv*) *If* $d = \infty$, *then* $\lambda_k \nearrow \infty$.

PROOF. (*i*), (*iv*) Follow from Proposition 3.52.

(*ii*) The first equality holds since $\mathcal{F}_1$ contains all antipodal pairs of points in $\mathcal{M}$ by Proposition 2.14 (*i*), and the second follows from homogeneity.

(*iii*) Follows from Proposition 3.53. □

The spectrum $\sigma(A_p, B_p)$ may possibly contain points other than those of the sequence (λ_k), and the eigenvalues λ_k may possibly be different from the standard ones

$$\lambda_k^\pm = \inf_{\substack{M \in \mathcal{F} \\ \gamma^\pm(M) \geqslant k}} \sup_{u \in M} \widetilde{\Psi}(u)$$

where

$$\gamma^+(M) := \sup \left\{k \geqslant 1 : \exists \text{ an odd continuous map } S^{k-1} \to M\right\}$$

and

$$\gamma^-(M) := \inf \left\{k \geqslant 1 : \exists \text{ an odd continuous map } M \to S^{k-1}\right\}$$

are the genus and the cogenus of $M \in \mathcal{F}$, respectively.

Proposition 4.7. *We have*

(*i*) $\lambda_1^\pm = \lambda_1$,

(*ii*) $\lambda_k^- \leqslant \lambda_k \leqslant \lambda_k^+, \quad k \geqslant 2$.

PROOF. (*i*) Since both the genus and the cogenus of an antipodal pair of points in $\mathcal{M}$ is 1,

$$\lambda_1^\pm = \min_{u \in \mathcal{M}} \widetilde{\Psi}(u) = \lambda_1$$

by Theorem 4.6 (*ii*).

(*ii*) If $M \in \mathcal{F}$ and there are odd continuous maps $S^{k^+-1} \to M \to S^{k^--1}$, then

$$k^+ \leqslant i(M) \leqslant k^-$$

by (i_2) and Example 2.11, so

$$\gamma^+(M) \leqslant i(M) \leqslant \gamma^-(M). \quad \square$$

Remarks 4.8. The sequence $\left(\lambda_k^-\right)$ is the more standard one. For the p-Laplacian, the sequence $\left(\lambda_k^+\right)$ was used to construct homotopically linking sets by Drábek and Robinson [**47**]. The sequence (λ_k) was introduced and used to obtain a nontrivial critical group by Perera [**98**]. It has also been used to construct homologically linking sets by Cingolani and Degiovanni [**30**] and homotopically linking sets via the piercing property by Perera and Szulkin [**105**].

4.3. Nontrivial Critical Groups

The functional associated with the eigenvalue problem (4.1) is

$$\Phi_\lambda(u) = I_p(u) - \lambda\, J_p(u), \quad u \in W. \tag{4.10}$$

When $\lambda \notin \sigma(A_p\,, B_p\,)$, the origin is the only critical point of Φ_λ and taking $U = W$ in (3.22) gives

$$C^q(\Phi_\lambda, 0) = H^q(\Phi_\lambda^0, \Phi_\lambda^0 \setminus \{0\}). \tag{4.11}$$

By homogeneity, Φ_λ^0 (radially) contracts to $\{0\}$ via

$$\Phi_\lambda^0 \times [0, 1] \to \Phi_\lambda^0, \quad (u, t) \mapsto (1 - t)\, u$$

and $\Phi_\lambda^0 \setminus \{0\}$ deformation retracts to $\Phi_\lambda^0 \cap \mathcal{M}$ via

$$(\Phi_\lambda^0 \setminus \{0\}) \times [0, 1] \to \Phi_\lambda^0 \setminus \{0\}\,, \quad (u, t) \mapsto (1 - t)\, u + t\, \pi_{\mathcal{M}}(u)$$

where

$$\pi_{\mathcal{M}} : W \setminus \{0\} \to \mathcal{M}, \quad u \mapsto \frac{u}{I_p(u)^{1/p}} \tag{4.12}$$

is the radial projection onto $\mathcal{M}$. Thus,

$$C^q(\Phi_\lambda, 0) \approx \begin{cases} \delta_{q0}\, \mathbb{Z}_2, & \Phi_\lambda^0 \cap \mathcal{M} = \emptyset \\ \widetilde{H}^{q-1}(\Phi_\lambda^0 \cap \mathcal{M}), & \Phi_\lambda^0 \cap \mathcal{M} \neq \emptyset \end{cases} \tag{4.13}$$

by Proposition 2.4.

Theorem 4.9. *Assume* (A_1) - (A_4), (B_1) - (B_3), *and* $\lambda \notin \sigma(A_p\,, B_p\,)$.

(*i*) *If* $\lambda < \lambda_1$, *then*

$$C^q(\Phi_\lambda, 0) \approx \delta_{q0}\, \mathbb{Z}_2.$$

(*ii*) *If* $\lambda > \lambda_1$, *then*

$$C^q(\Phi_\lambda, 0) \approx \widetilde{H}^{q-1}(\widetilde{\Psi}^\lambda) \quad \forall q.$$

In particular, $C^0(\Phi_\lambda, 0) = 0$.

(*iii*) *If* $\lambda_k < \lambda < \lambda_{k+1}$, *then*

$$C^k(\Phi_\lambda, 0) \neq 0.$$

PROOF. For $u \in \mathcal{M}$,

$$\Phi_\lambda(u) = 1 - \frac{\lambda}{\widetilde{\Psi}(u)}, \tag{4.14}$$

so $\Phi_\lambda^0 \cap \mathcal{M} = \widetilde{\Psi}^\lambda$. Since $\widetilde{\Psi}^\lambda = \emptyset$ if and only if $\lambda < \lambda_1$ by Theorem 4.6 (*ii*), (*i*) and (*ii*) follow from (4.13). Since $i(\widetilde{\Psi}^\lambda) = k$ when $\lambda_k < \lambda < \lambda_{k+1}$ by Theorem 4.6 (*iii*), then (*iii*) follows from Proposition 2.14 (*iv*). □

Corollary 4.10. *If* $\lambda \notin \sigma(A_p\,, B_p\,)$, *then* $C^q(\Phi_\lambda, 0) \neq 0$ *for some* q.

Remark 4.11. Theorem 4.9 was proved by Perera [**98**], where the sequence of eigenvalues (λ_k) was introduced precisely to obtain the nontrivial critical group in Corollary 4.10.

We will use the notations introduced in this chapter throughout the rest of the text.

CHAPTER 5

Existence Theory

In this chapter we study the global behavior of the functional Φ and obtain solutions of equation (1.1), together with some information about their critical groups, under various assumptions on the nonlinearity f at infinity. We assume that Φ has only a finite number of critical points.

We classify (1.1) according to the growth of the potential F as

(*i*) p-sublinear if

$$\lim_{t \to \infty} \frac{F(tu)}{t^p} \leqslant 0 \quad \forall u \neq 0, \tag{5.1}$$

(*ii*) asymptotically p-linear if

$$0 < \lim_{t \to \infty} \frac{F(tu)}{t^p} < \infty \quad \forall u \neq 0, \tag{5.2}$$

(*iii*) p-superlinear if

$$\lim_{t \to \infty} \frac{F(tu)}{t^p} = \infty \quad \forall u \neq 0. \tag{5.3}$$

5.1. p-Sublinear Case

If we strengthen (5.1) to

$$\lim_{\|u\| \to \infty} \frac{F(u)}{\|u\|^p} = 0,$$

then

$$\Phi(u) \geqslant \left(\frac{c_0}{p} + \mathrm{o}(1) \right) \|u\|^p,$$

so Φ is bounded from below and coercive, i.e.,

$$\lim_{\|u\| \to \infty} \Phi(u) = \infty.$$

More generally, we have

Theorem 5.1. *Assume* (A_1) - (A_4), (B_1) - (B_3), *and*

$$F(u) \leqslant \lambda \, J_p(u) + C \quad \forall u \in W \tag{5.4}$$

for some $\lambda < \lambda_1$ *and* $C > 0$. *Then*

(*i*) Φ *is coercive and satisfies* (PS),

(*ii*) Φ *is bounded from below,* $C^q(\Phi, \infty) \approx \delta_{q0} \, \mathbb{Z}_2$, *and* Φ *has a global minimizer* u *with* $C^q(\Phi, u) \approx \delta_{q0} \, \mathbb{Z}_2$.

PROOF. (*i*) Since $J_p \geqslant 0$, replacing λ with $\max\{\lambda, 0\}$ if necessary, we may assume that $\lambda \geqslant 0$, so

$$\begin{aligned}
(5.5) \qquad \Phi(u) &\geqslant I_p(u) - \lambda\, J_p(u) - C && \text{by (5.4)}\\
&\geqslant \left(1 - \frac{\lambda}{\lambda_1}\right) \frac{c_0}{p}\, \|u\|^p - C && \text{by Theorem 4.6 } (ii) \text{ and (1.6)}\\
&\to \infty \text{ as } \|u\| \to \infty.
\end{aligned}$$

Every (PS) sequence is bounded by coercivity and hence has a convergent subsequence by Lemma 3.3.

(*ii*) $\Phi \geqslant -C$ by (5.5), and then the rest follows from Corollary 3.19. □

Example 5.2. In Example 4.4, (5.4) holds if

$$F(x, t) \leqslant \frac{\lambda}{p}\, |t|^p + C \quad \forall (x, t)$$

for some $\lambda < \lambda_1$ and $C > 0$.

5.2. Asymptotically p-Linear Case

We strengthen (5.2) by assuming that

$$f = \lambda\, B_p - g \tag{5.6}$$

for some $\lambda > 0$ and a compact potential operator $g : W \to W^*$ satisfying

$$g(u) = \mathrm{o}(\|u\|^{p-1}) \text{ as } \|u\| \to \infty. \tag{5.7}$$

Then

$$\Phi(u) = \Phi_\lambda(u) + G(u) \tag{5.8}$$

where Φ_λ is given by (4.10) and

$$G = \lambda\, J_p - F \tag{5.9}$$

is the potential of g satisfying $G(0) = 0$. By Proposition 1.2 and (5.7),

$$G(u) = \int_0^1 \langle g(tu), u\rangle\, dt = \mathrm{o}(\|u\|^p) \text{ as } \|u\| \to \infty. \tag{5.10}$$

Example 5.3. In problem (1.9), assume that

$$f(x, t) = \lambda\, |t|^{p-2}\, t - g(x, t)$$

for some $g \in \mathrm{Car}(\Omega \times \mathbb{R})$ satisfying

$$g(x, t) = \mathrm{o}(|t|^{p-1}) \text{ as } |t| \to \infty, \text{ uniformly in } x.$$

Then we have

$$\begin{cases} -\Delta_p\, u = \lambda\, |u|^{p-2}\, u - g(x, u) & \text{in } \Omega\\ \\ \qquad u = 0 & \text{on } \partial\Omega, \end{cases}$$

and (5.6) and (5.7) hold with

$$\langle g(u), v\rangle = \int_\Omega g(x,u)\, v, \qquad G(u) = \int_\Omega G(x,u)$$

where

$$G(x,t) = \int_0^t g(x,s)\, ds.$$

To verify the (PS) (resp. (C)) condition for Φ, it suffices to show that every (PS) (resp. (C)) sequence is bounded by Lemma 3.3, and the following lemma is useful for this purpose.

Lemma 5.4. *If $\Phi'(u_j) \to 0$ and $\rho_j := \|u_j\| \to \infty$, then a subsequence of $\widetilde{u}_j := u_j/\rho_j$ converges to an eigenvector $\widetilde{u}$ associated with λ.*

PROOF. We have $\|\widetilde{u}_j\| = 1$ for all j, so a renamed subsequence converges weakly to some $\widetilde{u}$ since W is reflexive, and $B_p\, \widetilde{u}_j$ converges in W^* for a further subsequence since B_p is compact. Then

$$\langle A_p\, \widetilde{u}_j, \widetilde{u}_j - \widetilde{u}\rangle = \lambda \left\langle B_p\, \widetilde{u}_j, \widetilde{u}_j - \widetilde{u}\right\rangle + \frac{\langle \Phi'(u_j) - g(u_j), \widetilde{u}_j - \widetilde{u}\rangle}{\rho_j^{p-1}} \to 0$$

by Lemma 3.4 and (5.7), so $\widetilde{u}_j \to \widetilde{u}$ for a subsequence by (A_4). Now passing to the limit in

$$A_p\, \widetilde{u}_j = \lambda\, B_p\, \widetilde{u}_j + \frac{\Phi'(u_j) - g(u_j)}{\rho_j^{p-1}}$$

gives $A_p\, \widetilde{u} = \lambda\, B_p\, \widetilde{u}$, and $\widetilde{u} \neq 0$ since $\|\widetilde{u}\| = 1$. □

Since (5.9) and (5.10) imply that (5.4) holds with a slightly larger λ, here we assume that $\lambda_k \leqslant \lambda \leqslant \lambda_{k+1}$ for some $k \in \mathcal{K}$ such that $\lambda_k < \lambda_{k+1}$. Let $\mathcal{M}$ and $\widetilde{\Psi}$ be as in Section 4.1 and let $A_0 = \widetilde{\Psi}^{\lambda_k}$ and $B_0 = \widetilde{\Psi}_{\lambda_{k+1}}$. Then

$$i(A_0) = i(\mathcal{M} \backslash B_0) = k$$

by Theorem 4.6 (*iii*), so $A = RA_0$ cohomologically links $B = \pi_{\mathcal{M}}^{-1}(B_0) \cup \{0\}$ in dimension $k-1$ for any $R > 0$ by Proposition 3.26. By (5.8), (4.14), and (1.6),

$$\Phi(Ru) \leqslant G(Ru) - \left(\frac{\lambda}{\lambda_k} - 1\right) R^p, \quad u \in A_0 \tag{5.11}$$

and

$$\Phi(u) \geqslant \left(1 - \frac{\lambda}{\lambda_{k+1}}\right) \frac{c_0}{p}\, \|u\|^p + G(u), \quad u \in B. \tag{5.12}$$

Lemma 5.5. *If Φ satisfies* (C),

$$\lim_{R\to\infty} \left[G(Ru) - \left(\frac{\lambda}{\lambda_k} - 1\right) R^p\right] = -\infty, \quad \textit{uniformly in } u \in A_0, \tag{5.13}$$

and

$$\inf_{u \in B} \left[\left(1 - \frac{\lambda}{\lambda_{k+1}} \right) \frac{c_0}{p} \|u\|^p + G(u) \right] > -\infty, \tag{5.14}$$

then $C^0(\Phi, \infty) = 0$, $C^k(\Phi, \infty) \neq 0$, *and* (1.1) *has a solution* u *with* $C^k(\Phi, u) \neq 0$.

PROOF. Φ is bounded from below on B by (5.12) and (5.14), and for any $a < \inf \Phi(B)$ and sufficiently large R, (3.31) holds by (5.11) and (5.13), so the conclusion follows from Proposition 3.25. □

Lemma 5.6. *If* $\lambda > \lambda_k$, *then* (5.13) *holds, and if* $\lambda < \lambda_{k+1}$, *then* (5.14) *holds.*

PROOF. The first assertion follows since

$$\lim_{R \to \infty} \frac{G(Ru)}{R^p} = 0, \quad \text{uniformly in } u \in A_0$$

by (5.10) and (4.5), and the second follows similarly. □

Nonresonance. We say that (1.1) is nonresonant if $\lambda \notin \sigma(A_p, B_p)$ in (5.6).

Theorem 5.7. *Assume* (A_1) - (A_4), (B_1) - (B_3), *and* (5.6) *and* (5.7) *with* $\lambda \in (\lambda_k, \lambda_{k+1}) \backslash \sigma(A_p, B_p)$. *Then*

(*i*) Φ *satisfies* (PS),

(*ii*) $C^0(\Phi, \infty) = 0$, $C^k(\Phi, \infty) \neq 0$, *and* (1.1) *has a solution* u *with* $C^k(\Phi, u) \neq 0$.

PROOF. (*i*) Since $\lambda \notin \sigma(A_p, B_p)$, every (PS) sequence is bounded by Lemma 5.4 and hence has a convergent subsequence by Lemma 3.3.

(*ii*) Follows from Lemmas 5.6 and 5.5. □

Resonance. We say that (1.1) is resonant if $\lambda \in \sigma(A_p, B_p)$. Then we make an additional assumption on the non-p-homogeneous part of Φ given by

$$H(u) = \Phi(u) - \frac{1}{p} \langle \Phi'(u), u \rangle = G(u) - \frac{1}{p} \langle g(u), u \rangle$$

in order to ensure that Φ satisfies the (C) condition. Note that $(H(u_j))$ is bounded for every (C) sequence (u_j). Denoting by $\mathcal{N}$ the class of sequences $(u_j) \subset W$ such that $\rho_j := \|u_j\| \to \infty$ and $\widetilde{u}_j := u_j / \rho_j$ converges weakly to some $\widetilde{u} \neq 0$, we assume one of

$(H_\pm)$ H is bounded from below (resp. above) and every sequence $(u_j) \in \mathcal{N}$ has a subsequence such that

$$H(tu_j) \to \pm\infty \quad \forall t \geqslant 1.$$

In particular, no (C) sequence can belong to $\mathcal{N}$.

Lemma 5.8. *If* $(H_\pm)$ *holds, then* G *is bounded from below (resp. above) and every sequence* $(u_j) \in \mathcal{N}$ *has a subsequence such that*

$$G(u_j) \to \pm\infty. \tag{5.15}$$

PROOF. We have

$$\frac{d}{dt}\left(-\frac{G(tu)}{t^p}\right) = \frac{p\,H(tu)}{t^{p+1}},$$

and

$$\lim_{t\to\infty} \frac{G(tu)}{t^p} = 0$$

by (5.10), so

$$G(u) = \int_1^\infty \frac{p\,H(tu)}{t^{p+1}}\, dt.$$

Since

$$\int_1^\infty \frac{p}{t^{p+1}}\, dt = 1,$$

$\inf H \leqslant G \leqslant \sup H$ and (5.15) for the subsequence in $(H_\pm)$ follows from Fatou's lemma. □

We can now prove

Theorem 5.9. *Assume* (A_1) - (A_4), (B_1) - (B_3), *and* (5.6) *and* (5.7). *Then*

(*i*) Φ *satisfies* (C) *if* (H_+) *or* (H_-) *holds,*

(*ii*) Φ *is bounded from below,* $C^q(\Phi, \infty) \approx \delta_{q0}\,\mathbb{Z}_2$, *and* Φ *has a global minimizer* u *with* $C^q(\Phi, u) \approx \delta_{q0}\,\mathbb{Z}_2$ *if* $\lambda = \lambda_1$ *and* (H_+) *holds,*

(*iii*) $C^0(\Phi, \infty) = 0$, $C^k(\Phi, \infty) \neq 0$, *and* (1.1) *has a solution* u *with* $C^k(\Phi, u) \neq 0$ *in the following cases:*

(*a*) $\lambda \in [\lambda_k, \lambda_{k+1})$ *and* (H_-) *holds,*

(*b*) $\lambda \in (\lambda_k, \lambda_{k+1}]$ *and* (H_+) *holds.*

PROOF. (*i*) If a (C) sequence is unbounded, then Lemma 5.4 gives a subsequence that belongs to $\mathcal{N}$, contradicting $(H_\pm)$.

(*ii*) $\Phi = \Phi_{\lambda_1} + G$ is bounded from below since $\Phi_{\lambda_1} = I_p - \lambda_1 J_p \geqslant 0$ by Theorem 4.6 (*ii*) and G is bounded from below by Lemma 5.8, and then the rest follows from Corollary 3.19.

(*iii*) We apply Lemma 5.5.

(*a*) Since $\lambda < \lambda_{k+1}$, (5.14) holds by Lemma (5.6), and since $\lambda \geqslant \lambda_k$, it suffices to show that

$$\lim_{R\to\infty} G(Ru) = -\infty, \quad \text{uniformly in } u \in A_0$$

to verify (5.13). If not, there are sequences $(u_j) \subset A_0$ and $R_j \to \infty$ such that $(G(R_j\, u_j))$ is bounded from below. Then $\rho_j := R_j \|u_j\| \to \infty$ by (4.5), a renamed subsequence of $\widetilde{u}_j := R_j\, u_j/\rho_j = u_j/\|u_j\|$ converges weakly to some

$\widetilde{u}$ since W is reflexive, and $B_p \widetilde{u}_j$ converges to some $L \in W^*$ for a further subsequence since B_p is compact. We have

$$\begin{aligned}
\langle L, \widetilde{u} \rangle &= \lim \langle B_p \widetilde{u}_j, \widetilde{u}_j \rangle && \text{by Lemma 3.4}\\
&= \lim \frac{p}{\widetilde{\Psi}(u_j)\, \|u_j\|^p}\\
&\geqslant \frac{c_0}{\lambda_k} && \text{by (4.5)}\\
&> 0 && \text{by Theorem 4.6 } (ii)
\end{aligned}$$

and hence $\widetilde{u} \neq 0$, so $(R_j\, u_j) \in \mathcal{N}$. But, then $G(R_j\, u_j) \to -\infty$ for a subsequence by Lemma 5.8, a contradiction.

(*b*) Since $\lambda > \lambda_k$, (5.13) holds by Lemma (5.6), and since $\lambda \leqslant \lambda_{k+1}$ and G is bounded from below by Lemma 5.8, (5.14) holds as well. □

Example 5.10. In Example 5.3,

$$H(u) = \int_\Omega H(x, u)$$

where

$$H(x, t) = G(x, t) - \frac{1}{p}\, g(x, t)\, t$$

is the non-p-homogeneous part of $G(x, t)$. We claim that $(H_\pm)$ holds if

$$H(x, t) \geqslant \text{ (resp. } \leqslant) \; C(x) \quad \text{a.e.} \tag{5.16}$$

for some $C \in L^1(\Omega)$ and

$$H(x, t) \to \pm\infty \quad \text{a.e. as } |t| \to \infty.$$

By (5.16), H is bounded from below (resp. above). If $(u_j) \in \mathcal{N}$, then for a subsequence, $\widetilde{u}_j \to \widetilde{u}$ a.e. and hence

$$H(tu_j) \geqslant \text{ (resp. } \leqslant) \int_{\widetilde{u}(x) \neq 0} H(x, t\rho_j\, \widetilde{u}_j(x)) + \int_{\widetilde{u}(x)=0} C(x) \to \pm\infty \quad \forall t \geqslant 1$$

by Fatou's lemma.

Remarks 5.11. The conditions on $H(x, t)$ in Example 5.10 were given by Perera [**97**]. Different conditions for the existence of a solution, without the information on its nontrivial critical group, were given by Drábek and Robinson [**47**].

5.3. p-Superlinear Case

In addition to (5.3), we assume that F is bounded from below and, in order to ensure that Φ satisfies the (PS) condition,

$$H_\mu(u) := F(u) - \frac{1}{\mu}\, \langle f(u), u \rangle$$

is bounded from above for some $\mu > p$.

For any u on the unit sphere S,

$$\begin{aligned}
\frac{d}{dt}\left(\Phi(tu)\right) &= p\,t^{p-1}\,I_p(u) - \langle f(tu), u\rangle \\
&= \frac{1}{t}\left[p\,\Phi(tu) - (\mu - p)\,F(tu) + \mu\,H_\mu(tu)\right] \\
&\leqslant \frac{p}{t}\left(\Phi(tu) - a_0\right) \qquad (5.17)
\end{aligned}$$

where

$$a_0 := \inf\, \frac{1}{p}\left[(\mu - p)\,F - \mu\,H_\mu\right] > -\infty,$$

so all critical values of Φ are greater than or equal to a_0. Note that $a_0 \leqslant 0$ since $F(0) = H_\mu(0) = 0$.

By (5.3),

$$\Phi(tu) = t^p\left(I_p(u) - \frac{F(tu)}{t^p}\right) \to -\infty \text{ as } t \to \infty \qquad (5.18)$$

for any $u \in S$. The following lemma describes the structure of the sublevel sets at infinity.

Lemma 5.12. *For each $a < a_0$, there is a C^1-map $T_a : S \to (0, \infty)$ such that*

$$\Phi^a = \left\{tu : u \in S,\ t \geqslant T_a(u)\right\} \simeq S.$$

PROOF. We have $\Phi(tu) \leqslant a$ for all sufficiently large $t > 0$ by (5.18), and

$$\Phi(tu) \leqslant a \implies \frac{d}{dt}\left(\Phi(tu)\right) < 0$$

by (5.17), so there is a unique $T_a(u) > 0$ such that

$$t < \text{ (resp. } =, >)\ T_a(u) \implies \Phi(tu) > \text{ (resp. } =, <)\ a,$$

and the map T_a is C^1 by the implicit function theorem. Then $W\setminus\{0\}$, which is $\simeq S$, radially deformation retracts to $\Phi^a = \left\{tu : u \in S,\ t \geqslant T_a(u)\right\}$ via

$$(W\setminus\{0\}) \times [0,1] \to W\setminus\{0\},$$

$$(u,t) \mapsto \begin{cases} (1-t)\,u + t\,T_a(\pi_S(u))\,\pi_S(u), & u \in (W\setminus\{0\})\setminus\Phi^a \\ u, & u \in \Phi^a. \end{cases} \qquad \square$$

We can now prove

Theorem 5.13. *Assume* (A_1) - (A_4), *F is bounded from below and satisfies* (5.3), *and H_μ is bounded from above for some $\mu > p$. Then*

(*i*) *Φ satisfies* (PS),
(*ii*) $C^q(\Phi, \infty) \approx \delta_{qd}\,\mathbb{Z}_2$,
(*iii*) *Φ is bounded from above and has a global maximizer u with $C^q(\Phi, u) \approx \delta_{qd}\,\mathbb{Z}_2$ if $d < \infty$.*

PROOF. (*i*) If (u_j) is a (PS) sequence, then (1.6) gives

$$\begin{aligned} c_0 \left(\frac{1}{p} - \frac{1}{\mu}\right) \|u_j\|^p &\leqslant \left(1 - \frac{p}{\mu}\right) I_p(u_j) \\ &= \Phi(u_j) - \frac{1}{\mu} \big\langle \Phi'(u_j), u_j \big\rangle + H_\mu(u_j) \\ &\leqslant \mathrm{o}(\|u_j\|) + \mathrm{O}(1), \end{aligned}$$

so (u_j) is bounded and hence has a convergent subsequence by Lemma 3.3.

(*ii*) Since any $a < a_0$ is less than all critical values and Φ is unbounded from below by (5.18),

$$\begin{aligned} C^q(\Phi, \infty) &\approx \widetilde{H}^{q-1}(\Phi^a) && \text{by Proposition 3.15 } (ii) \\ &\approx \widetilde{H}^{q-1}(S) && \text{by Lemma 5.12} \\ &\approx \delta_{qd}\, \mathbb{Z}_2 && \text{by Example 2.5.} \end{aligned}$$

(*iii*) Since S is compact, so is $\Phi_a = \big\{tu : u \in S,\ 0 \leqslant t \leqslant T_a(u)\big\}$ and hence Φ is bounded there. Since $\Phi < a$ outside Φ_a, then Φ is bounded from above everywhere, and the rest follows from Propositions 3.17 and 3.18. □

Example 5.14. In problem (1.9), we claim that all the hypotheses of Theorem 5.13 are satisfied if

$$0 < F(x,t) \leqslant \frac{1}{\mu}\, f(x,t)\, t, \quad |t| \geqslant T \tag{5.19}$$

for some $\mu > p$ and $T > 0$. By (1.10), $f(x,t)$, and hence also $F(x,t)$, is bounded on bounded t intervals. Integrating (5.19) gives

$$F(x,t) \geqslant c(x)\, |t|^\mu - C \quad \forall (x,t)$$

where $c(x) = \min F(x, \pm T)/T^\mu > 0$ and $C > 0$, so

$$\frac{F(tu)}{t^p} \geqslant t^{\mu-p} \int_\Omega c(x)\, |u|^\mu - \frac{C\,|\Omega|}{t^p} \to \infty \text{ as } t \to \infty \quad \forall u \neq 0$$

and $F(u) \geqslant -C\,|\Omega|$ where $|\Omega|$ is the volume of Ω. Finally, since

$$H_\mu(x,t) := F(x,t) - \frac{1}{\mu}\, f(x,t)\, t$$

is $\leqslant 0$ for $|t| \geqslant T$ by (5.19) and bounded for $|t| < T$,

$$H_\mu(u) = \int_\Omega H_\mu(x,u)$$

is bounded from above.

Remarks 5.15. In the semilinear case $p = 2$ of Example 5.14, condition (5.19) was introduced by Ambrosetti and Rabinowitz [**7**] and the critical groups at infinity were determined by Wang [**132**].

CHAPTER 6

Monotonicity and Uniqueness

In this short chapter we give simple sufficient conditions for equation (1.1) to have at most one solution and show how they can be verified in applications.

Definition 6.1. An operator $T : W \to W^*$ is monotone (resp. strictly monotone) if

$$\langle T(u) - T(v), u - v \rangle \geqslant \text{ (resp. } >) \, 0 \quad \forall u \neq v.$$

Theorem 6.2. *If A_p is strictly monotone and $-f$ is monotone, then* (1.1) *has at most one solution.*

PROOF. If both u_1 and u_2 are solutions, then

$$\langle A_p \, u_1 - A_p \, u_2, u_1 - u_2 \rangle = \langle f(u_1) - f(u_2), u_1 - u_2 \rangle \leqslant 0$$

and hence $u_1 = u_2$. □

The following lemma is useful for verifying the strict monotonicity of A_p.

Lemma 6.3. *If*

$$\langle A_p \, u, v \rangle \leqslant \|u\|^{p-1} \, \|v\| \quad \forall u, v \in W$$

and the equality holds if and only if $\alpha u = \beta v$ for some $\alpha, \beta \geqslant 0$, not both zero, then A_p is strictly monotone.

PROOF. As in (1.8),

$$\langle A_p \, u - A_p \, v, u - v \rangle \geqslant \big(\|u\|^{p-1} - \|v\|^{p-1} \big) \big(\|u\| - \|v\| \big) \geqslant 0.$$

If the leftmost term is zero, equality holds throughout and hence $\langle A_p \, u, v \rangle = \|u\|^{p-1} \, \|v\|$, $\langle A_p \, v, u \rangle = \|v\|^{p-1} \, \|u\|$, and $\|u\| = \|v\|$. Then $\alpha u = \beta v$ for some $\alpha, \beta \geqslant 0$, not both zero, so either $u = v = 0$, or $\alpha = \beta > 0$ and hence $u = v \neq 0$. □

Example 6.4. In problem (1.9), $-f$ is clearly monotone if $f(x, t)$ is nonincreasing in t, and now we show that the strict monotonicity of A_p follows from Lemma 6.3. By Schwarz and Hölder inequalities,

$$\langle A_p \, u, v \rangle = \int_\Omega |\nabla u|^{p-2} \, \nabla u \cdot \nabla v \leqslant \int_\Omega |\nabla u|^{p-1} \, |\nabla v| \leqslant \|u\|^{p-1} \, \|v\| \, .$$

Clearly, equality holds throughout if $\alpha u = \beta v$ for some $\alpha, \beta \geqslant 0$, not both zero. Conversely, if $\langle A_p\, u, v\rangle = \|u\|^{p-1}\,\|v\|$, equality holds in both inequalities. The equality in the Hölder inequality gives

$$\alpha\,|\nabla u| = \beta\,|\nabla v| \;\text{ a.e.}$$

for some $\alpha, \beta \geqslant 0$, not both zero, and then the equality in the Schwarz inequality gives

$$\alpha\,\nabla u = \beta\,\nabla v \;\text{ a.e.,}$$

so $\alpha u = \beta v$.

CHAPTER 7

Nontrivial Solutions and Multiplicity

In many applications $f(0) = 0$ and hence equation (1.1) has the trivial solution $u = 0$, and we are interested in finding others. Throughout this chapter we assume that this is the case and there is only a finite number of solutions.

7.1. Mountain Pass Solutions

First we obtain nontrivial mountain pass solutions of (1.1) in the asymptotically p-linear and p-superlinear cases assuming that either

$$F(u) \leqslant \lambda_1 \, J_p(u) \quad \forall u \in B_\rho(0) \tag{7.1}$$

for some $\rho > 0$, or

$$F(u) \leqslant \overline{\lambda} \, J_p(u) + \mathrm{o}(\|u\|^p) \;\text{ as }\; u \to 0 \tag{7.2}$$

for some $0 \leqslant \overline{\lambda} < \lambda_1$.

Lemma 7.1. *If* (7.1) *or* (7.2) *holds, then the origin is a local minimizer of* Φ *and hence* $C^q(\Phi, 0) \approx \delta_{q0} \, \mathbb{Z}_2$.

Proof. Since $I_p(u)/J_p(u) \geqslant \lambda_1$ for all $u \neq 0$ by Theorem 4.6 (ii), if (7.1) holds, then

$$\Phi(u) \geqslant I_p(u) - \lambda_1 \, J_p(u) \geqslant 0 = \Phi(0) \quad \forall u \in B_\rho(0),$$

and if (7.2) holds, then

$$\frac{\Phi(u)}{I_p(u)} \geqslant 1 - \frac{\overline{\lambda}}{\lambda_1} + \mathrm{o}(1) \;\text{ as }\; u \to 0$$

by (1.6). Proposition 3.18 gives the critical groups. □

Theorem 7.2. *Assume* (A_1) - (A_4), (B_1) - (B_3), (5.6) *and* (5.7), *and* (7.1) *or* (7.2). *Then* (1.1) *has a mountain pass solution* $u \neq 0$ *with* $\Phi(u) > 0$ *in the following cases:*

(a) $\lambda \in (\lambda_k, \lambda_{k+1}) \backslash \sigma(A_p, B_p)$,
(b) $\lambda \in [\lambda_k, \lambda_{k+1})$ *and* (H_-) *holds,*
(c) $\lambda \in (\lambda_k, \lambda_{k+1}]$ *and* (H_+) *holds.*

Proof. Follows from Corollary 3.31 since $C^0(\Phi, 0) \approx \mathbb{Z}_2$ by Lemma 7.1 and $C^0(\Phi, \infty) = 0$ and Φ satisfies (C) by Theorem 5.7 in case (a) and Theorem 5.9 in cases (b) and (c). □

Theorem 7.3. *Assume* (A_1) - (A_4), (B_1) - (B_3), F *is bounded from below and satisfies* (5.3) *and* (7.1) *or* (7.2), *and* H_μ *is bounded from above for some* $\mu > p$. *Then* (1.1) *has a mountain pass solution* $u_1 \neq 0$ *with* $\Phi(u_1) > 0$. *If* $2 \leqslant d < \infty$, *then there is a second solution* $u_2 \neq 0$ *that is a global maximizer of* Φ *with* $\Phi(u_2) > 0$.

PROOF. Since $C^0(\Phi, 0) \approx \mathbb{Z}_2$ by Lemma 7.1 and $C^0(\Phi, \infty) = 0$ and Φ satisfies (PS) by Theorem 5.13, Φ has a mountain pass point $u_1 \neq 0$ with $\Phi(u_1) > 0$ by Corollary 3.31. When $2 \leqslant d < \infty$, there is a global maximizer u_2 with $C^q(\Phi, u_2) \approx \delta_{qd}\, \mathbb{Z}_2$ by Theorem 5.13 *(iii)*, $\Phi(u_2) \geqslant \Phi(u_1) > 0$, and $u_2 \neq u_1$ since $C^1(\Phi, u_2) \neq C^1(\Phi, u_1)$. □

Example 7.4. When $p > n$ in Example 4.4, (7.1) holds if

$$F(x, t) \leqslant \frac{\lambda_1}{p}\, |t|^p, \quad |t| \leqslant \delta \tag{7.3}$$

for some $\delta > 0$ by the Sobolev imbedding $W^{1,p}_0(\Omega) \hookrightarrow C(\overline{\Omega})$. When $p \leqslant n$, we strengthen (7.3) to

$$F(x, t) \leqslant \frac{\overline{\lambda}}{p}\, |t|^p, \quad |t| \leqslant \delta$$

for some $0 \leqslant \overline{\lambda} < \lambda_1$. Then

$$F(x, t) \leqslant \frac{\overline{\lambda}}{p}\, |t|^p + C\, |t|^r \quad \forall (x, t)$$

for some $r \in (p, p^*)$ and $C > 0$ by (1.10), so (7.2) holds by the imbedding $W^{1,p}_0(\Omega) \hookrightarrow L^r(\Omega)$.

7.2. Solutions via a Cohomological Local Splitting

Next we obtain nontrivial solutions of (1.1) in the p-sublinear and p-super-linear cases assuming that either

$$\lambda_k\, J_p(u) \leqslant F(u) < \lambda_{k+1}\, J_p(u) \quad \forall u \in B_\rho(0) \setminus \{0\} \tag{7.4}$$

for some $k \in \mathcal{K}$ such that $\lambda_k < \lambda_{k+1}$ and $\rho > 0$, or

$$\underline{\lambda}\, J_p(u) + \mathrm{o}(\|u\|^p) \leqslant F(u) \leqslant \overline{\lambda}\, J_p(u) + \mathrm{o}(\|u\|^p) \ \text{ as } \ u \to 0 \tag{7.5}$$

for some $\lambda_k < \underline{\lambda} \leqslant \overline{\lambda} < \lambda_{k+1}$.

Lemma 7.5. *If* (7.4) *or* (7.5) *holds, then* Φ *has a cohomological local splitting near zero in dimension* k *and hence* $C^k(\Phi, 0) \neq 0$.

PROOF. We take $U = rI\mathcal{M}$, $h(u) = ru$, $A_0 = \widetilde{\Psi}^{\lambda_k}$, and $B_0 = \widetilde{\Psi}_{\lambda_{k+1}}$ with $\mathcal{M}$ and $\widetilde{\Psi}$ as in Section 4.1 and $r > 0$ sufficiently small. Theorem 4.6 *(iii)* gives (3.32) with $q = k$. If (7.4) holds, taking $r \leqslant (c_0/p)^{1/p}\, \rho$, we have $U \subset B_\rho(0)$ by (4.5) and hence

$$1 - \frac{\lambda_{k+1}}{\widetilde{\Psi}(\pi_{\mathcal{M}}(u))} < \frac{\Phi(u)}{I_p(u)} \leqslant 1 - \frac{\lambda_k}{\widetilde{\Psi}(\pi_{\mathcal{M}}(u))} \quad \forall u \in U \setminus \{0\},$$

which implies (3.33). If (7.5) holds, then

$$1 - \frac{\overline{\lambda}}{\widetilde{\Psi}(\pi_{\mathcal{M}}(u))} + \mathrm{o}(1) \leqslant \frac{\Phi(u)}{I_p(u)} \leqslant 1 - \frac{\underline{\lambda}}{\widetilde{\Psi}(\pi_{\mathcal{M}}(u))} + \mathrm{o}(1) \quad \text{as } u \to 0$$

by (1.6), which also implies (3.33) when r is sufficiently small. Taking r smaller if necessary, we may assume that U contains no other critical points. Proposition 3.34 gives the nontrivial critical group. □

Theorem 7.6. *Assume* (A_1) - (A_4), (B_1) - (B_3), (5.4), *and* (7.4) *or* (7.5). *Then* (1.1) *has a solution* $u_1 \neq 0$. *If* $k \geqslant 2$, *then there is a second solution* $u_2 \neq 0$.

Proof. Follows from Corollary 3.32 since $C^k(\Phi, 0) \neq 0$ by Lemma 7.5 and Φ is bounded from below and satisfies (PS) by Theorem 5.1. □

Theorem 7.7. *Assume* (A_1) - (A_4), (B_1) - (B_3), F *is bounded from below and satisfies* (5.3) *and* (7.4) *or* (7.5) *with* $k \leqslant d - 1$, *and* H_μ *is bounded from above for some* $\mu > p$. *Then* (1.1) *has a solution* $u_1 \neq 0$ *with either* $\Phi(u_1) < 0$ *and* $C^{k-1}(\Phi, u_1) \neq 0$, *or* $\Phi(u_1) > 0$ *and* $C^{k+1}(\Phi, u_1) \neq 0$. *If* $k + 2 \leqslant d < \infty$, *then there is a second solution* $u_2 \neq 0$ *that is a global maximizer of* Φ *with* $\Phi(u_2) > 0$.

Proof. Since $C^k(\Phi, 0) \neq 0$ by Lemma 7.5 and $C^k(\Phi, \infty) = 0$ and Φ satisfies (PS) by Theorem 5.13, Φ has a critical point $u_1 \neq 0$ with either $\Phi(u_1) < 0$ and $C^{k-1}(\Phi, u_1) \neq 0$, or $\Phi(u_1) > 0$ and $C^{k+1}(\Phi, u_1) \neq 0$ by Proposition 3.28 *(ii)*. When $k + 2 \leqslant d < \infty$, there is a global maximizer u_2 with $C^q(\Phi, u_2) \approx \delta_{qd}\, \mathbb{Z}_2$ by Theorem 5.13 *(iii)*, $\Phi(u_2) > \Phi(0) = 0$ since $C^k(\Phi, 0) \neq 0$ and hence the origin is not a global maximizer by Proposition 3.18, and $u_2 \neq u_1$ since $C^{k-1}(\Phi, u_2) = C^{k+1}(\Phi, u_2) = 0$. □

Example 7.8. When $p > n$ in Example 4.4, (7.4) holds if

$$\frac{\lambda_k}{p}\, |t|^p \leqslant F(x, t) < \frac{\lambda_{k+1}}{p}\, |t|^p, \quad 0 < |t| \leqslant \delta$$

for some $\delta > 0$, and when $p \leqslant n$, (7.5) holds if

$$\frac{\underline{\lambda}}{p}\, |t|^p \leqslant F(x, t) \leqslant \frac{\overline{\lambda}}{p}\, |t|^p, \quad |t| \leqslant \delta$$

for some $\lambda_k < \underline{\lambda} \leqslant \overline{\lambda} < \lambda_{k+1}$.

7.3. Nonlinearities that Cross an Eigenvalue

Now we obtain nontrivial solutions of (1.1) in the asymptotically p-linear case assuming that

$$F(u) \geqslant \lambda_k\, J_p(u) \quad \forall u \in W \tag{7.6}$$

for some $k \in \mathcal{K}$ such that $\lambda_k < \lambda_{k+1}$. Let $A_0 = \widetilde{\Psi}^{\lambda_k}$,

$$\mathcal{H} = \big\{h \in C(CA_0, \mathcal{M}) : h(CA_0) \text{ is closed in } \mathcal{M},\ h|_{A_0} = id_{A_0}\big\},$$

and

$$\underline{\lambda} = \inf_{h \in \mathcal{H}} \sup_{u \in h(CA_0)} \widetilde{\Psi}(u).$$

Each $h \in \mathcal{H}$ extends to an odd map $\widetilde{h} \in C(SA_0, \mathcal{M})$ since $h|_{A_0} = id_{A_0}$ and A_0 is symmetric, and $\sup \widetilde{\Psi}(h(CA_0)) = \sup \widetilde{\Psi}(\widetilde{h}(SA_0))$ since $\widetilde{\Psi}$ is even. By (i_2), (i_6), and Theorem 4.6 (iii),

$$i(\widetilde{h}(SA_0)) \geqslant i(SA_0) = i(A_0) + 1 = k + 1,$$

so $\widetilde{h}(SA_0) \in \mathcal{F}_{k+1}$ and hence $\sup \widetilde{\Psi}(\widetilde{h}(SA_0)) \geqslant \lambda_{k+1}$. It follows that $\underline{\lambda} \geqslant \lambda_{k+1}$.

Theorem 7.9. *Assume* (A_1) - (A_4), (B_1) - (B_3), (5.6) *and* (5.7) *with* $\lambda \notin \sigma(A_p, B_p)$, *and* (7.6). *If* $\lambda > \underline{\lambda}$ *and either*

$$F(u) \leqslant \lambda_{k+1} J_p(u) \quad \forall u \in B_\rho(0) \tag{7.7}$$

for some $\rho > 0$, *or*

$$F(u) \leqslant \overline{\lambda} J_p(u) + o(\|u\|^p) \ \text{ as } u \to 0 \tag{7.8}$$

for some $\overline{\lambda} < \lambda_{k+1}$, *then* (1.1) *has a solution* $u \neq 0$ *with* $\Phi(u) \geqslant 0$.

Proof. By (7.6),

$$\Phi(tu) \leqslant -\left(\frac{\lambda_k}{\widetilde{\Psi}(u)} - 1\right) t^p \leqslant 0, \quad u \in A_0,\ t \geqslant 0.$$

Since $\lambda > \underline{\lambda}$, there is a $h \in \mathcal{H}$ such that $\lambda' := \sup \widetilde{\Psi}(h(CA_0)) < \lambda$. By (5.8) and (4.14),

$$\Phi(Ru) \leqslant -\left(\frac{\lambda}{\lambda'} - 1 - \frac{G(Ru)}{R^p}\right) R^p, \quad u \in h(CA_0),\ R > 0,$$

and by (5.10) and (4.5),

$$\frac{G(Ru)}{R^p} \to 0 \text{ as } R \to \infty, \text{ uniformly in } u \in h(CA_0).$$

It follows that $\Phi \leqslant 0$ on $A = R(IA_0 \cup h(CA_0))$ for all sufficiently large R.

Let $B_0 = \widetilde{\Psi}_{\lambda_{k+1}}$. Then $\Phi \geqslant 0$ on $B = rB_0$ when $r > 0$ is sufficiently small. Indeed, if (7.7) holds, taking $r = (c_0/p)^{1/p} \rho$, we have $B \subset B_\rho(0)$ by (4.5) and hence

$$\Phi(ru) \geqslant \left(1 - \frac{\lambda_{k+1}}{\widetilde{\Psi}(u)}\right) r^p \geqslant 0 \quad \forall\, ru \in B,$$

and if (7.8) holds, then

$$\Phi(ru) \geqslant \left(1 - \frac{\overline{\lambda}}{\lambda_{k+1}} + \mathrm{o}(1)\right) r^p \text{ as } r \to 0,\ u \in B_0.$$

Since

$$i(A_0) = i(\mathcal{M} \backslash B_0) = k$$

by Theorem 4.6 (*iii*), A homotopically links B when $R > r$ by Proposition 3.23, and since $\lambda \notin \sigma(A_p, B_p)$, Φ satisfies (PS) by Theorem 5.7. Applying Proposition 3.21, $c \geqslant \inf \Phi(B) \geqslant 0$ defined by (3.27) with $g : A \subset W$ and $X = IA$ is a critical value of Φ, and there is a corresponding critical point on B when $c = 0$. □

Example 7.10. In Example 4.4, (7.6) holds if

$$F(x,t) \geqslant \frac{\lambda_k}{p}\,|t|^p \quad \forall (x,t).$$

Remarks 7.11. Theorem 7.9 in the p-Laplacian case is due to Perera and Szulkin [**105**]. The semilinear case $p = 2$ is a well-known result of Amann and Zehnder [**4**].

7.4. Odd Nonlinearities

Finally we assume that f is odd, so that Φ is even, and obtain multiple solutions of (1.1) using Propositions 3.38 and 3.44.

Theorem 7.12. *Assume* (A_1) - (A_4), (B_1) - (B_3), (5.4), *and f is odd. If*

$$F(u) \geqslant \underline{\lambda}\, J_p(u) + o(\|u\|^p) \quad \text{as } u \to 0 \tag{7.9}$$

for some $\underline{\lambda} > \lambda_m$, then (1.1) *has m distinct pairs of solutions at negative levels.*

Proof. We apply Proposition 3.38, taking $A = r\,\widetilde{\Psi}^{\lambda_m}$ with $r > 0$ sufficiently small and $B = W$. We have $i(A) \geqslant m$ by Theorem 4.6 (*iii*) and $i(B^c) = 0$ by (i_1). By (7.9) and (4.5),

$$\Phi(ru) \leqslant -\left(\frac{\underline{\lambda}}{\lambda_m} - 1 + \mathrm{o}(1)\right) r^p \quad \text{as } r \to 0,\ u \in \widetilde{\Psi}^{\lambda_m},$$

so $\sup \Phi(A) < 0$ when r is sufficiently small. By Theorem 5.1, Φ is bounded from below, and hence $\inf \Phi(B) > -\infty$, and satisfies (PS). The conclusion follows. □

When (1.1) is asymptotically p-linear we only consider the nonresonant case for the sake of simplicity.

Theorem 7.13. *Assume* (A_1) - (A_4), (B_1) - (B_3), (5.6) *and* (5.7) *with* $\lambda \notin \sigma(A_p, B_p)$, *and f is odd.*

(*i*) *If $\lambda < \lambda_k$ and*

$$F(u) \geqslant \underline{\lambda}\, J_p(u) + o(\|u\|^p) \quad \text{as } u \to 0 \tag{7.10}$$

for some $\underline{\lambda} > \lambda_{k+m-1}$, then (1.1) *has m distinct pairs of solutions at negative levels.*

(*ii*) *If $\lambda > \lambda_{k+m-1}$ and*

$$F(u) \leqslant \overline{\lambda}\, J_p(u) + o(\|u\|^p) \quad \text{as } u \to 0 \tag{7.11}$$

for some $\overline{\lambda} < \lambda_k$, then (1.1) *has m distinct pairs of solutions at positive levels.*

PROOF. Since $\lambda \notin \sigma(A_p\,, B_p\,)$, Φ satisfies (PS) by Theorem 5.7.

(*i*) We apply Proposition 3.38, taking $A = r\,\widetilde{\Psi}^{\lambda_{k+m-1}}$ with $r > 0$ sufficiently small and $B = \pi_{\mathcal{M}}^{-1}(\widetilde{\Psi}_{\lambda_k}) \cup \{0\}$. We have $i(A) \geqslant k + m - 1$ and

$$i(B^c) = i(\pi_{\mathcal{M}}^{-1}(\mathcal{M}\backslash\widetilde{\Psi}_{\lambda_k})) = i(\mathcal{M}\backslash\widetilde{\Psi}_{\lambda_k}) \leqslant k - 1$$

by Theorem 4.6 (*iii*). By (7.10) and (4.5),

$$\Phi(ru) \leqslant -\left(\frac{\lambda}{\lambda_{k+m-1}} - 1 + \mathrm{o}(1)\right) r^p \text{ as } r \to 0,\ u \in \widetilde{\Psi}^{\lambda_{k+m-1}},$$

so $\sup \Phi(A) < 0$ when r is sufficiently small. By (5.8), (4.14), (1.6), and (5.10),

$$\Phi(u) \geqslant \left[\left(1 - \frac{\lambda}{\lambda_k}\right)\frac{c_0}{p} + \mathrm{o}(1)\right] \|u\|^p \quad \text{as} \quad \|u\| \to \infty,\ u \in B,$$

so $\inf \Phi(B) > -\infty$.

(*ii*) By Proposition 2.14 (*iii*), $\mathcal{M}\backslash\widetilde{\Psi}_\lambda$ has a compact subset $C \in \mathcal{F}$ with $i(C) = i(\mathcal{M}\backslash\widetilde{\Psi}_\lambda)$. We apply Proposition 3.44, taking $A = RC$ and $B = r\,\widetilde{\Psi}_{\lambda_k} \subset r\mathcal{M}$ with $r > 0$ sufficiently small and $R > r$ sufficiently large. We have

$$i(A) = i(C) = i(\mathcal{M}\backslash\widetilde{\Psi}_\lambda) \geqslant i(\widetilde{\Psi}^{\lambda_{k+m-1}}) \geqslant k + m - 1$$

and $i(r\mathcal{M}\backslash B) \leqslant k - 1$ by Theorem 4.6 (*iii*). By (5.8) and (4.14),

$$\Phi(Ru) \leqslant -\left(\frac{\lambda}{\max \widetilde{\Psi}(C)} - 1 - \frac{G(Ru)}{R^p}\right) R^p, \quad u \in C,\ R > 0,$$

and by (5.10) and (4.5),

$$\frac{G(Ru)}{R^p} \to 0 \text{ as } R \to \infty, \text{ uniformly in } u \in C.$$

Since $\max \widetilde{\Psi}(C) < \lambda$ by the continuity of $\widetilde{\Psi}$, it follows that $\sup \Phi(A) \leqslant 0$ when R is sufficiently large. By (7.11) and (4.5),

$$\Phi(ru) \geqslant \left(1 - \frac{\overline{\lambda}}{\lambda_k} + \mathrm{o}(1)\right) r^p \text{ as } r \to 0,\ u \in \widetilde{\Psi}_{\lambda_k},$$

so $\inf \Phi(B) > 0$ when r is sufficiently small. □

Theorem 7.14. *Assume* (A_1) - (A_4), (B_1) - (B_3), F *is bounded from below and satisfies* (5.3), H_μ *is bounded from above for some* $\mu > p$, *and* f *is odd. If*

$$F(u) \leqslant \overline{\lambda}\, J_p(u) + o(\|u\|^p) \text{ as } u \to 0$$

for some $\overline{\lambda} < \lambda_k$, *then* (1.1) *has* $d - k + 1$ *distinct pairs of solutions at positive levels. If* $d = \infty$, *then there is an unbounded sequence of positive critical levels.*

PROOF. We will show that Φ has m distinct pairs of critical points in $\Phi^{-1}(0,\infty)$ for any $m \leqslant d-k+1$ in $\mathbb{N}$. We apply Proposition 3.44, noting that Φ satisfies (PS) by Theorem 5.13, and taking A to be the intersection of $S_R(0)$ with any $(k+m-1)$-dimensional subspace of W and $B = r\,\widetilde{\Psi}_{\lambda_k} \subset r\mathcal{M}$, with $r > 0$ sufficiently small and $R > (p/c_0)^{1/p}\, r$ sufficiently large. A is a compact symmetric set with index $k+m-1$ by (i_8) and $i(r\mathcal{M}\backslash B) \leqslant k-1$ by Theorem 4.6 *(iii)*. By (5.18), $\sup \Phi(A) \leqslant 0$ when R is sufficiently large, and as in the proof of Theorem 7.13 *(ii)*, $\inf \Phi(B) > 0$ when r is sufficiently small.

When $d = \infty$, the sequence of critical levels in $(0,\infty)$ given by Proposition 3.44 is unbounded by Proposition 3.42 *(ii)*. □

Remark 7.15. Theorem 7.13 in the p-Laplacian case is due to Perera and Szulkin [**105**].

CHAPTER 8

Jumping Nonlinearities and the Dancer-Fučík Spectrum

Consider the problem

$$A_p\, u = a\, B_p^+\, u + b\, B_p^-\, u \tag{8.1}$$

in W^*, where A_p satisfies (A_1) - (A_4), $B_p^\pm : W \to W^*$ are

$(B_1^\pm)$ $(p-1)$-homogeneous,
$(B_2^\pm)$ nonnegative:

$$\langle B_p^\pm\, u, u\rangle \geqslant 0 \quad \forall u \in W,$$

$(B_3^\pm)$ compact potential operators,
$(B_4^\pm)$ $B_p^+ + B_p^- = B_p$,

and $a, b \in \mathbb{R}$. The set $\Sigma(A_p, B_p^+, B_p^-)$ of all points $(a, b) \in \mathbb{R}^2$ such that there is a $u \neq 0$ in W satisfying (8.1) is called the Dancer-Fučík spectrum of the triple of operators (A_p, B_p^+, B_p^-).

By Proposition 1.2, the potentials $J_p^\pm$ of $B_p^\pm$ satisfying $J_p^\pm(0) = 0$ are given by

$$J_p^\pm(u) = \frac{1}{p}\,\langle B_p^\pm\, u, u\rangle$$

and are p-homogeneous. By $(B_2^\pm)$, $J_p^\pm(u) \geqslant 0$ for all $u \in W$. Moreover,

$$J_p^+(u) + J_p^-(u) = J_p(u) > 0 \quad \forall u \neq 0 \tag{8.2}$$

by $(B_4^\pm)$ and (B_2), so either $J_p^+(u) > 0$ or $J_p^-(u) > 0$ when $u \neq 0$. If $(a, b) \in \Sigma(A_p, B_p^+, B_p^-)$ and $u \neq 0$ satisfies (8.1), applying it to u and using (8.2) gives

$$I_p(u) = a\, J_p^+(u) + b\, J_p^-(u) \leqslant \max\{a, b\}\; J_p(u)$$

and hence $\max\{a, b\} \geqslant \lambda_1$ by Theorem 4.6 *(ii)*, so

$$\Sigma(A_p, B_p^+, B_p^-) \subset \big\{(a, b) \in \mathbb{R}^2 : \max\{a, b\} \geqslant \lambda_1\big\}.$$

By $(B_4^\pm)$, $\Sigma(A_p, B_p^+, B_p^-)$ contains the set $\big\{(\lambda, \lambda) : \lambda \in \sigma(A_p, B_p)\big\}$, in particular, the sequence of points $((\lambda_k, \lambda_k))_{k \in \mathcal{K}}$.

Example 8.1. In Example 4.1, the usual choices are

$$\langle B_p^\pm\, u, v\rangle = \pm \int_\Omega (u^\pm)^{p-1}\, v$$

where $u^{\pm} = \max\{\pm u, 0\}$ are the positive and negative parts of u, respectively:

$$\begin{cases} -\Delta_p u = a\,(u^+)^{p-1} - b\,(u^-)^{p-1} & \text{in } \Omega \\ \\ u = 0 & \text{on } \partial\Omega. \end{cases}$$

The Dancer-Fučík spectrum Σ_p was introduced in the semilinear case $p = 2$ of this problem by Dancer [**38, 39**] and Fučík [**51**], who recognized its significance for the solvability of (1.9) when

$$f(x,t) = a\,t^+ - b\,t^- + \mathrm{o}(|t|) \text{ as } |t| \to \infty, \text{ uniformly in } x.$$

In the ODE case $n = 1$, Fučík showed that Σ_2 consists of a sequence of hyperbolic like curves passing through the points (λ_k, λ_k), with one or two curves going through each point, and Drábek [**46**] has recently shown that Σ_p has this same general shape for all p in this case. In the PDE case $n \geqslant 2$, Σ_2 consists locally of curves emanating from the points (λ_k, λ_k), in particular, contains two strictly decreasing curves, which may coincide, such that the points in the square $(\lambda_{k-1}, \lambda_{k+1}) \times (\lambda_{k-1}, \lambda_{k+1})$ that are either below the lower curve or above the upper curve are not in Σ_2, while the points between them may or may not belong to Σ_2 when they do not coincide (see Các [**22**], Cuesta and Gossez [**37**], de Figueiredo and Gossez [**41**], Margulies and Margulies [**77**], and Schechter [**119**]). Since φ_1 solves

$$\begin{cases} -\Delta_p u = \lambda_1\,(u^+)^{p-1} - b\,(u^-)^{p-1} & \text{in } \Omega \\ \\ u = 0 & \text{on } \partial\Omega \end{cases}$$

for arbitrary b and $-\varphi_1$ solves

$$\begin{cases} -\Delta_p u = a\,(u^+)^{p-1} - \lambda_1\,(u^-)^{p-1} & \text{in } \Omega \\ \\ u = 0 & \text{on } \partial\Omega \end{cases}$$

for arbitrary a, Σ_p contains the two lines $\lambda_1 \times \mathbb{R}$ and $\mathbb{R} \times \lambda_1$. A first nontrivial curve in Σ_p passing through (λ_2, λ_2) and asymptotic to $\lambda_1 \times \mathbb{R}$ and $\mathbb{R} \times \lambda_1$ at infinity was constructed using a mountain pass scheme by Cuesta, de Figueiredo, and Gossez [**36**]. More recently, unbounded sequences of decreasing curves of Σ_p, analogous to the lower and upper curves of Schechter in the semilinear case, have been constructed using various minimax schemes by Cuesta [**35**], Micheletti and Pistoia [**83**], and Perera [**101**].

We close this introductory section with

Proposition 8.2. *The spectrum $\Sigma(A_p\,, B_p^+\,, B_p^-)$ is closed.*

PROOF. Let the sequence $((a_j, b_j)) \subset \Sigma(A_p\,, B_p^+\,, B_p^-)$ converge to $(a, b) \in \mathbb{R}^2$ and let $u_j \neq 0$ satisfy

$$A_p\,u_j = a_j\,B_p^+\,u_j + b_j\,B_p^-\,u_j. \tag{8.3}$$

Using (A_1) and $(B_1^\pm)$ to replace u_j with $u_j/\|u_j\|$ if necessary, we may assume that $\|u_j\| = 1$, so a renamed subsequence converges weakly to some u since W is reflexive, and $B_p^\pm\, u_j$ converge in W^* for further subsequences since $B_p^\pm$ are compact. Then

$$\langle A_p\, u_j, u_j - u\rangle = a_j \left\langle B_p^+\, u_j, u_j - u\right\rangle + b_j \left\langle B_p^-\, u_j, u_j - u\right\rangle \to 0$$

by Lemma 3.4, so $u_j \to u$ for a subsequence by (A_4). Now passing to the limit in (8.3) shows that u satisfies (8.1), and $u \neq 0$ since $\|u\| = 1$, so $(a, b) \in \Sigma(A_p\,, B_p^+\,, B_p^-\,)$. □

8.1. Variational Setting

For each $s > 0$,

$$B_p^s := B_p^+ + s\, B_p^-$$

is a $(p-1)$-homogeneous compact potential operator from W to W^* by $(B_1^\pm)$ and $(B_3^\pm)$, and the p-homogeneous potential J_p^s of B_p^s satisfying $J_p^s(0) = 0$ is given by

$$J_p^s(u) = \frac{1}{p} \left\langle B_p^s\, u, u\right\rangle = J_p^+(u) + s\, J_p^-(u).$$

Since either $J_p^+(u) > 0$ or $J_p^-(u) > 0$ when $u \neq 0$, $J_p^s(u) > 0$ for all $u \neq 0$, so the functional

$$\Psi_s(u) = \frac{1}{J_p^s(u)}, \qquad u \in W \backslash \{0\}$$

is positive and its restriction

$$\widetilde{\Psi}_s = \Psi_s|_{\mathcal{M}}$$

to the manifold $\mathcal{M}$ defined by (4.4) is C^1. We will show that the points of $\Sigma(A_p\,, B_p^+\,, B_p^-\,)$ that are on the ray $b = sa$, $a \geqslant 0$ are of the form (c, sc) with c a critical value of $\widetilde{\Psi}_s$.

Since

$$\Psi_s'(u) = -\widetilde{\Psi}_s(u)^2\, B_p^s\, u,$$

it follows from Proposition 3.54 that the norm of $\widetilde{\Psi}_s'(u) \in T_u^*\mathcal{M}$ is given by

$$\left\|\widetilde{\Psi}_s'(u)\right\|_u^* = \min_{\mu \in \mathbb{R}} \left\|\mu\, A_p\, u + \widetilde{\Psi}_s(u)^2\, B_p^s\, u\right\|^*. \tag{8.4}$$

Lemma 8.3. *The point $(c, sc) \in \Sigma(A_p\,, B_p^+\,, B_p^-\,)$ if and only if c is a critical value of $\widetilde{\Psi}_s$.*

Proof. By (8.4), $\widetilde{\Psi}_s'(u) = 0$ if and only if

$$\mu\, A_p\, u + \widetilde{\Psi}_s(u)^2\, B_p^s\, u = 0 \tag{8.5}$$

for some $\mu \in \mathbb{R}$. If (8.5) holds, applying it to u gives

$$\mu = -\widetilde{\Psi}_s(u)^2\, \frac{\left\langle B_p^s\, u, u\right\rangle}{\langle A_p\, u, u\rangle} = -\widetilde{\Psi}_s(u)^2\, \frac{J_p^s(u)}{I_p(u)} = -\widetilde{\Psi}_s(u) < 0,$$

so (8.5) reduces to

$$A_p\, u = c\, B_p^s\, u \tag{8.6}$$

where $c = \widetilde{\Psi}_s(u)$. Conversely, if $u \in \mathcal{M}$ solves (8.6),

$$c = \frac{I_p(u)}{J_p^s(u)} = \widetilde{\Psi}_s(u),$$

so (8.6) implies (8.5) with $\mu = -\widetilde{\Psi}_s(u)$. □

Example 8.4. In Example 8.1,

$$J_p^{\pm}(u) = \frac{1}{p} \int_\Omega (u^{\pm})^p, \qquad \widetilde{\Psi}_s(u) = \frac{p}{\displaystyle\int_\Omega (u^+)^p + s\,(u^-)^p}.$$

Lemma 8.5. *$\widetilde{\Psi}_s$ satisfies* (PS).

Proof. Let $(u_j) \subset \mathcal{M}$ be a $(\mathrm{PS})_c$ sequence. By (4.5), (u_j) is bounded, so a renamed subsequence converges weakly to some $u \in W$ since W is reflexive, and $B_p^s\, u_j$ converges to some $L \in W^*$ for a further subsequence since B_p^s is compact. Then

$$\widetilde{\Psi}_s(u_j) = \frac{p}{\big\langle B_p^s\, u_j, u_j \big\rangle} \to \frac{p}{\langle L, u \rangle} \neq 0$$

by Lemma 3.4, so $c \neq 0$.

By (8.4),

$$\mu_j\, A_p\, u_j + \widetilde{\Psi}_s(u_j)^2\, B_p^s\, u_j \to 0 \tag{8.7}$$

for some sequence $(\mu_j) \subset \mathbb{R}$. Applying (8.7) to u_j gives $\mu_j + \widetilde{\Psi}_s(u_j) \to 0$, so $\mu_j \to -c \neq 0$. Now applying (8.7) to $u_j - u$ gives $\langle A_p\, u_j, u_j - u \rangle \to 0$ since $\big\langle B_p^s\, u_j, u_j - u \big\rangle \to 0$ by Lemma 3.4, so $u_j \to u \in \mathcal{M}$ for a subsequence by (A_4). □

8.2. A Family of Curves in the Spectrum

We now construct an unbounded sequence of decreasing (continuous) curves in $\Sigma(A_p\,, B_p^+\,, B_p^-\,)$.

Lemma 8.6. *If $s_1 \leqslant s_2$, then*

$$\widetilde{\Psi}_{s_1} \geqslant \widetilde{\Psi}_{s_2}, \qquad s_1\, \widetilde{\Psi}_{s_1} \leqslant s_2\, \widetilde{\Psi}_{s_2}.$$

Proof. We have

$$\widetilde{\Psi}_{s_1} - \widetilde{\Psi}_{s_2} = (s_2 - s_1)\, J_p^-\, \widetilde{\Psi}_{s_1}\, \widetilde{\Psi}_{s_2} \geqslant 0$$

and

$$s_1\, \widetilde{\Psi}_{s_1} - s_2\, \widetilde{\Psi}_{s_2} = (s_1 - s_2)\, J_p^+\, \widetilde{\Psi}_{s_1}\, \widetilde{\Psi}_{s_2} \leqslant 0. \quad \square$$

First we construct a curve $\mathcal{C}_1$ passing through the point (λ_1, λ_1). Let

$$c_1(s) = \inf_{u \in \mathcal{M}} \widetilde{\Psi}_s(u)$$

and

$$\mathcal{C}_1 = \big\{(c_1(s), s\, c_1(s)) : s > 0\big\}.$$

Theorem 8.7. *Assume* (A_1) - (A_4), (B_1) - (B_3), *and* $(B_1^\pm)$ - $(B_4^\pm)$. *Then* $\mathcal{C}_1$ *is a decreasing curve in* $\Sigma(A_p\,, B_p^+\,, B_p^-\,)$ *passing through* (λ_1, λ_1).

PROOF. $c_1(s)$ is a critical value of $\widetilde{\Psi}_s$ by Proposition 3.48 and hence $\mathcal{C}_1 \subset \Sigma(A_p\,, B_p^+\,, B_p^-\,)$ by Lemma 8.3. If $s_1 \leqslant s_2$, then

$$c_1(s_1) \geqslant c_1(s_2), \qquad s_1\, c_1(s_1) \leqslant s_2\, c_1(s_2) \tag{8.8}$$

by Lemma 8.6, so $\mathcal{C}_1$ is decreasing. To see that it is continuous, note that (8.8) implies

$$c_1(s_2) \leqslant c_1(s_1) \leqslant \frac{s_2}{s_1}\, c_1(s_2), \qquad \frac{s_1}{s_2}\, c_1(s_1) \leqslant c_1(s_2) \leqslant c_1(s_1)$$

and hence $c_1(s_1) \to c_1(s_2)$ as $s_1 \nearrow s_2$ and $c_1(s_2) \to c_1(s_1)$ as $s_2 \searrow s_1$. Since $J_p^+ + J_p^- = J_p$ by (8.2), $\widetilde{\Psi}_1$ is the functional $\widetilde{\Psi}$ defined by (4.6), so $c_1(1) = \lambda_1$ by Theorem 4.6 (ii) and hence $\mathcal{C}_1$ passes through (λ_1, λ_1). □

Now we construct a curve $\mathcal{C}_k$ in the spectrum for each $k \geqslant 2$ in $\mathcal{K}$ such that $\lambda_k > \lambda_{k-1}$. Let $A_k = \widetilde{\Psi}^{\lambda_{k-1}}$,

$$\Gamma_k = \big\{\gamma \in C(CA_k, \mathcal{M}) : \gamma|_{A_k} = id_{A_k}\big\},$$

$$c_k(s) = \inf_{\gamma \in \Gamma_k} \sup_{u \in \gamma(CA_k)} \widetilde{\Psi}_s(u),$$

and

$$\mathcal{C}_k = \big\{(c_k(s), s\, c_k(s)) : \lambda_{k-1}/\lambda_k < s < \lambda_k/\lambda_{k-1}\big\}.$$

Theorem 8.8. *Assume* (A_1) - (A_4), (B_1) - (B_3), *and* $(B_1^\pm)$ - $(B_4^\pm)$. *Then* $\mathcal{C}_k$ *is a decreasing curve in* $\Sigma(A_p\,, B_p^+\,, B_p^-\,)$ *and* $c_k(1) \geqslant \lambda_k$.

PROOF. Let $B_k = \widetilde{\Psi}_{\lambda_k}$. Then

$$i(A_k) = i(\mathcal{M} \backslash B_k) = k - 1$$

by Theorem 4.6 (iii) and hence A_k homotopically links B_k by Proposition 3.49. Since

$$\frac{\widetilde{\Psi}}{\max\{s, 1\}} \leqslant \widetilde{\Psi}_s \leqslant \frac{\widetilde{\Psi}}{\min\{s, 1\}},$$

when $\lambda_{k-1}/\lambda_k < s < \lambda_k/\lambda_{k-1}$,

$$\sup \widetilde{\Psi}_s(A_k) \leqslant \frac{\lambda_{k-1}}{\min\{s, 1\}} < \frac{\lambda_k}{\max\{s, 1\}} \leqslant \inf \widetilde{\Psi}_s(B_k).$$

It follows from Proposition 3.50 that

$$c_k(s) \geqslant \frac{\lambda_k}{\max\{s, 1\}}$$

is a critical value of $\widetilde{\Psi}_s$ and hence $\mathcal{C}_k \subset \Sigma(A_p\,, B_p^+\,, B_p^-\,)$. By Lemma 8.6,

$$c_k(s_1) \geqslant c_k(s_2), \qquad s_1\, c_k(s_1) \leqslant s_2\, c_k(s_2) \quad \forall s_1 \leqslant s_2,$$

and the monotonicity and continuity of $\mathcal{C}_k$ follow as in the proof of Theorem 8.7. □

8.3. Homotopy Invariance of Critical Groups

The functional associated with problem (8.1) is

$$\Phi_{(a,b)}(u) = I_p(u) - J_p^{(a,b)}(u), \quad u \in W \tag{8.9}$$

where $J_p^{(a,b)} = a\, J_p^+ + b\, J_p^-$ is the potential of $B_p^{(a,b)} = a\, B_p^+ + b\, B_p^-$ satisfying $J_p^{(a,b)}(0) = 0$, and the origin is the only critical point of $\Phi_{(a,b)}$ when $(a,b) \notin \Sigma(A_p\,, B_p^+\,, B_p^-\,)$. We will show that the critical groups $C^q(\Phi_{(a,b)}, 0)$ are constant on path components of $\mathbb{R}^2 \backslash \Sigma(A_p\,, B_p^+\,, B_p^-\,)$.

Taking $U = W$ in (3.22) gives

$$C^q(\Phi_{(a,b)}, 0) = H^q(\Phi^0_{(a,b)}, \Phi^0_{(a,b)} \backslash \{0\}),$$

and for $u \neq 0$,

$$\Phi_{(a,b)}(u) = \left(1 - \widetilde{J}_p^{(a,b)}(\pi_{\mathcal{M}}(u))\right) I_p(u) \tag{8.10}$$

where $\widetilde{J}_p^{(a,b)} = J_p^{(a,b)}\Big|_{\mathcal{M}}$ and $\pi_{\mathcal{M}}$ is the radial projection onto $\mathcal{M}$ given by (4.12).

Lemma 8.9. *If $(a,b) \notin \Sigma(A_p\,, B_p^+\,, B_p^-\,)$, then*

$$C^q(\Phi_{(a,b)}, 0) \approx \begin{cases} \delta_{q0}\, \mathbb{Z}_2, & (\widetilde{J}_p^{(a,b)})_1 = \emptyset \\ \widetilde{H}^{q-1}((\widetilde{J}_p^{(a,b)})_1), & (\widetilde{J}_p^{(a,b)})_1 \neq \emptyset. \end{cases}$$

PROOF. By homogeneity, $\Phi^0_{(a,b)}$ contracts to $\{0\}$ via

$$\Phi^0_{(a,b)} \times [0,1] \to \Phi^0_{(a,b)}, \quad (u,t) \mapsto (1-t)\, u$$

and $\Phi^0_{(a,b)} \backslash \{0\}$ deformation retracts to $\Phi^0_{(a,b)} \cap \mathcal{M}$ via

$$(\Phi^0_{(a,b)} \backslash \{0\}) \times [0,1] \to \Phi^0_{(a,b)} \backslash \{0\}\,, \quad (u,t) \mapsto (1-t)\, u + t\, \pi_{\mathcal{M}}(u).$$

Since $\Phi^0_{(a,b)} \cap \mathcal{M} = (\widetilde{J}_p^{(a,b)})_1$ by (8.10), the conclusion follows from Proposition 2.4. □

It follows from Proposition 3.54 that the norm of $(\widetilde{J}_p^{(a,b)})'(u) \in T_u^* \mathcal{M}$ is given by

$$\left\| (\widetilde{J}_p^{(a,b)})'(u) \right\|_u^* = \min_{\mu \in \mathbb{R}} \left\| \mu\, A_p\, u - B_p^{(a,b)}\, u \right\|^*. \tag{8.11}$$

Lemma 8.10. *The point $(a,b) \in \Sigma(A_p\,, B_p^+\,, B_p^-\,)$ if and only if 1 is a critical value of $\widetilde{J}_p^{(a,b)}$.*

PROOF. By (8.11), $(\widetilde{J}_p^{(a,b)})'(u) = 0$ if and only if

$$\mu\, A_p\, u = B_p^{(a,b)}\, u \tag{8.12}$$

for some $\mu \in \mathbb{R}$. Applying (8.12) to u gives $\mu = \widetilde{J}_p^{(a,b)}(u)$, so $u \in \mathcal{M}$ is a critical point of $\widetilde{J}_p^{(a,b)}$ with $\widetilde{J}_p^{(a,b)}(u) = 1$ if and only if (8.1) holds. □

The special case $(a_j, b_j) \equiv (a, b)$ of the following lemma shows that $\widetilde{J}_p^{(a,b)}$ satisfies $(\mathrm{PS})_c$ for all $c \neq 0$.

Lemma 8.11. *If $(a_j, b_j) \to (a, b)$ in $\mathbb{R}^2$, $(u_j) \subset \mathcal{M}$, and*

$$\widetilde{J}_p^{(a_j,b_j)}(u_j) \to c \neq 0, \quad \big\|(\widetilde{J}_p^{(a_j,b_j)})'(u_j)\big\|_{u_j}^* \to 0,$$

then a subsequence of (u_j) converges to a critical point of $\widetilde{J}_p^{(a,b)}$ with critical value c.

PROOF. By (4.5), (u_j) is bounded, so a renamed subsequence converges weakly to some $u \in W$ since W is reflexive, and $B_p^{\pm}\, u_j$ converge to some $L^{\pm} \in W^*$ for further subsequences since $B_p^{\pm}$ are compact. Then

$$B_p^{(a_j,b_j)}\, u_j = a_j\, B_p^+\, u_j + b_j\, B_p^-\, u_j \to a\, L^+ + b\, L^-. \tag{8.13}$$

By (8.11),

$$\mu_j\, A_p\, u_j - B_p^{(a_j,b_j)}\, u_j \to 0 \tag{8.14}$$

for some sequence $(\mu_j) \subset \mathbb{R}$. Applying (8.14) to u_j gives $\mu_j - \widetilde{J}_p^{(a_j,b_j)}(u_j) \to 0$, so $\mu_j \to c \neq 0$. Now applying (8.14) to $u_j - u$ gives $\langle A_p\, u_j, u_j - u\rangle \to 0$ since $\big\langle B_p^{(a_j,b_j)}\, u_j, u_j - u\big\rangle \to 0$ by (8.13) and Lemma 3.4, so $u_j \to u$ for a subsequence by (A_4). Then $u \in \mathcal{M}$ since $\mathcal{M}$ is closed, and $A_p\, u_j \to A_p\, u$ and $L^{\pm} = B_p^{\pm}\, u$ by the continuity of A_p and $B_p^{\pm}$, respectively. Hence $B_p^{(a_j,b_j)}\, u_j \to B_p^{(a,b)}\, u$ by (8.13) and

$$c\, A_p\, u - B_p^{(a,b)}\, u = 0 \tag{8.15}$$

by (8.14). So u is a critical point of $\widetilde{J}_p^{(a,b)}$ by (8.11), and applying (8.15) to u gives $\widetilde{J}_p^{(a,b)}(u) = c$. □

We are now ready to prove

Theorem 8.12. *Assume (A_1) - (A_4), (B_1) - (B_3), and $(B_1^{\pm})$ - $(B_4^{\pm})$. If there is a path in $\mathbb{R}^2 \backslash \Sigma(A_p\,, B_p^+\,, B_p^-\,)$ joining the points (a_0, b_0) and (a_1, b_1), then*

$$C^q(\Phi_{(a_0,b_0)}, 0) \approx C^q(\Phi_{(a_1,b_1)}, 0) \quad \forall q.$$

Since such a path is compact and $C^q(\Phi_{(a,b)}, 0)$ is determined by the homotopy type of $(\widetilde{J}_p^{(a,b)})_1$ by Lemma 8.9, it suffices to prove the following local version.

Theorem 8.13. *If* $(a,b) \in \mathbb{R}^2 \backslash \Sigma(A_p, B_p^+, B_p^-)$*, then there is a* $\delta > 0$ *such that*

$$(\widetilde{J}_p^{(a',b')})_1 \simeq (\widetilde{J}_p^{(a,b)})_1 \quad \forall (a',b') \in B_\delta(a,b).$$

Since the set $\mathbb{R}^2 \backslash \Sigma(A_p, B_p^+, B_p^-)$ is open by Proposition 8.2, for all sufficiently small $\delta > 0$, $B_\delta(a,b) \subset \mathbb{R}^2 \backslash \Sigma(A_p, B_p^+, B_p^-)$ and hence 1 is a regular value of $\widetilde{J}_p^{(a',b')}$ for all $(a',b') \in B_\delta(a,b)$ by Lemma 8.10. Lemma 8.11 gives the following stronger version.

Lemma 8.14. *If* $(a,b) \in \mathbb{R}^2 \backslash \Sigma(A_p, B_p^+, B_p^-)$*, then there is a* $\delta > 0$ *such that* $\widetilde{J}_p^{(a',b')}$ *has no critical values in* $[1, 1 + 3\delta/\lambda_1]$ *for all* $(a',b') \in B_\delta(a,b)$*.*

Proof. If not, there are sequences $\delta_j \searrow 0$, $(a_j', b_j') \in B_{\delta_j}(a,b)$, and $(u_j) \subset \mathcal{M}$ such that

$$1 \leqslant \widetilde{J}_p^{(a_j',b_j')}(u_j) \leqslant 1 + 3\delta_j/\lambda_1, \quad (\widetilde{J}_p^{(a_j',b_j')})'(u_j) = 0.$$

Then a subsequence of (u_j) converges to a critical point of $\widetilde{J}_p^{(a,b)}$ with critical value 1 by Lemma 8.11 and hence $(a,b) \in \Sigma(A_p, B_p^+, B_p^-)$ by Lemma 8.10, a contradiction. □

Proof of Theorem 8.13. Let $\delta > 0$ be as in Lemma 8.14 and let $(a',b') \in B_\delta(a,b)$. Then

$$\begin{aligned}\left|\widetilde{J}_p^{(a',b')}(u) - \widetilde{J}_p^{(a,b)}(u)\right| &\leqslant |a' - a|\, J_p^+(u) + |b' - b|\, J_p^-(u)\\ &\leqslant \delta\, J_p(u) \leqslant \delta/\lambda_1 \quad \forall u \in \mathcal{M}\end{aligned}$$

by Theorem 4.6 (ii), so

$$(\widetilde{J}_p^{(a,b)})_1 \supset (\widetilde{J}_p^{(a',b')})_{1+\delta/\lambda_1} \supset (\widetilde{J}_p^{(a,b)})_{1+2\delta/\lambda_1} \supset (\widetilde{J}_p^{(a',b')})_{1+3\delta/\lambda_1}.$$

Since $\widetilde{J}_p^{(a,b)}$ has no critical values in $[1, 1 + 2\delta/\lambda_1]$, there is a deformation retraction

$$\eta : (\widetilde{J}_p^{(a,b)})_1 \times [0,1] \to (\widetilde{J}_p^{(a,b)})_1$$

of $(\widetilde{J}_p^{(a,b)})_1$ onto $(\widetilde{J}_p^{(a,b)})_{1+2\delta/\lambda_1}$ by Lemma 3.47 applied to $-\widetilde{J}_p^{(a,b)}$. Similarly, there is a deformation retraction

$$\eta' : (\widetilde{J}_p^{(a',b')})_{1+\delta/\lambda_1} \times [0,1] \to (\widetilde{J}_p^{(a',b')})_{1+\delta/\lambda_1}$$

of $(\widetilde{J}_p^{(a',b')})_{1+\delta/\lambda_1}$ onto $(\widetilde{J}_p^{(a',b')})_{1+3\delta/\lambda_1}$. The map

$$(\widetilde{J}_p^{(a,b)})_1 \times [0,1] \to (\widetilde{J}_p^{(a,b)})_1,$$

$$(u,t) \mapsto \begin{cases} \eta(u, 2t), & 0 \leqslant t \leqslant 1/2 \\ \eta'(\eta(u,1), 2t-1), & 1/2 < t \leqslant 1 \end{cases}$$

is then a deformation retraction of $(\widetilde{J}_p^{(a,b)})_1$ onto $(\widetilde{J}_p^{(a',b')})_{1+3\delta/\lambda_1}$, and $(\widetilde{J}_p^{(a',b')})_{1+3\delta/\lambda_1}$ is also a deformation retract of $(\widetilde{J}_p^{(a',b')})_1$ since $\widetilde{J}_p^{(a',b')}$ has no critical values in $[1, 1+3\delta/\lambda_1]$, so

$$(\widetilde{J}_p^{(a',b')})_1 \simeq (\widetilde{J}_p^{(a',b')})_{1+3\delta/\lambda_1} \simeq (\widetilde{J}_p^{(a,b)})_1. \quad \square$$

Since $\Phi_{(\lambda,\lambda)}$ is the functional Φ_λ defined by (4.10), the following corollary is immediate from Theorem 8.12 and Corollary 4.10.

Corollary 8.15. *If (a,b) is in a path component of $\mathbb{R}^2 \backslash \Sigma(A_p, B_p^+, B_p^-)$ containing a point of the form (λ,λ), then $C^q(\Phi_{(a,b)}, 0) \neq 0$ for some q.*

Remark 8.16. Theorem 8.12 in the p-Laplacian case is due to Dancer and Perera [**40**].

8.4. Perturbations and Solvability

Now we consider the solvability of (1.1) when

$$f = a\, B_p^+ + b\, B_p^- - g = B_p^{(a,b)} - g \tag{8.16}$$

for some $(a,b) \in \mathbb{R}^2 \backslash \Sigma(A_p, B_p^+, B_p^-)$ and a compact potential operator $g : W \to W^*$ satisfying

$$g(u) = \mathrm{o}(\|u\|^{p-1}) \text{ as } \|u\| \to \infty. \tag{8.17}$$

Then

$$\Phi(u) = \Phi_{(a,b)}(u) + G(u)$$

where $\Phi_{(a,b)}$ is given by (8.9) and

$$G = a\, J_p^+ + b\, J_p^- - F = J_p^{(a,b)} - F$$

is the potential of g satisfying $G(0) = 0$. By Proposition 1.2 and (8.17),

$$G(u) = \int_0^1 \langle g(tu), u \rangle \, dt = \mathrm{o}(\|u\|^p) \text{ as } \|u\| \to \infty. \tag{8.18}$$

Example 8.17. In problem (1.9), assume that

$$f(x,t) = a\,(t^+)^{p-1} - b\,(t^-)^{p-1} - g(x,t)$$

for some $g \in \mathrm{Car}(\Omega \times \mathbb{R})$ satisfying

$$g(x,t) = \mathrm{o}(|t|^{p-1}) \text{ as } |t| \to \infty, \text{ uniformly in } x.$$

Then we have

$$\begin{cases} -\Delta_p\, u = a\,(u^+)^{p-1} - b\,(u^-)^{p-1} - g(x,u) & \text{in } \Omega \\ u = 0 & \text{on } \partial\Omega, \end{cases}$$

and (8.16) and (8.17) hold with

$$\langle g(u), v \rangle = \int_\Omega g(x,u)\, v, \qquad G(u) = \int_\Omega G(x,u)$$

where

$$G(x,t) = \int_0^t g(x,s)\,ds.$$

The following lemma, together with Lemma 3.3, shows that both Φ and $\Phi_{(a,b)}$ satisfy the (PS) condition when $(a,b) \notin \Sigma(A_p\,, B_p^+\,, B_p^-\,)$.

Lemma 8.18. *If* (8.17) *holds,* $\Phi'(u_j) \to 0$*, and* $\rho_j := \|u_j\| \to \infty$*, then a subsequence of* $\widetilde{u}_j := u_j/\rho_j$ *converges to a nontrivial solution* $\widetilde{u}$ *of* (8.1).

PROOF. We have $\|\widetilde{u}_j\| = 1$ for all j, so a renamed subsequence converges weakly to some $\widetilde{u}$ since W is reflexive, and $B_p^{(a,b)}\,\widetilde{u}_j$ converges in W^* for a further subsequence since $B_p^{(a,b)}$ is compact. Then

$$\langle A_p\,\widetilde{u}_j, \widetilde{u}_j - \widetilde{u}\rangle = \left\langle B_p^{(a,b)}\,\widetilde{u}_j, \widetilde{u}_j - \widetilde{u}\right\rangle + \frac{\langle \Phi'(u_j) - g(u_j), \widetilde{u}_j - \widetilde{u}\rangle}{\rho_j^{p-1}} \to 0$$

by Lemma 3.4 and (8.17), so $\widetilde{u}_j \to \widetilde{u}$ for a subsequence by (A_4). Now passing to the limit in

$$A_p\,\widetilde{u}_j = B_p^{(a,b)}\,\widetilde{u}_j + \frac{\Phi'(u_j) - g(u_j)}{\rho_j^{p-1}}$$

gives $A_p\,\widetilde{u} = B_p^{(a,b)}\,\widetilde{u}$, and $\widetilde{u} \neq 0$ since $\|\widetilde{u}\| = 1$. □

For $R > 0$, let

$$U_R = R I \mathcal{M} = \left\{u \in W : I_p(u) \leqslant R^p\right\}.$$

By (1.6), U_R is a bounded neighborhood of the origin with boundary $\partial U_R = R\mathcal{M}$. Since $(a,b) \notin \Sigma(A_p\,, B_p^+\,, B_p^-\,)$, Lemma 8.18 implies that the critical set of Φ lies in U_R if R is sufficiently large. The following lemma shows that outside U_{2R} the lower-order term G can be deformed away without changing the critical set.

Lemma 8.19. *If* $(a,b) \notin \Sigma(A_p\,, B_p^+\,, B_p^-\,)$*, then there are an* $R > 0$ *and a* C^1*-functional* Φ_R *such that*

(*i*) $\Phi_R(u) = \begin{cases} \Phi(u), & u \in U_R \\ \Phi_{(a,b)}(u), & u \in \overline{U_{2R}^c}, \end{cases}$

(*ii*) *all critical points of both* Φ *and* Φ_R *are in* $\mathring{U}_R$*, and hence the solutions of equation* (1.1) *coincide with the critical points of* Φ_R *also by* (*i*),

(*iii*) Φ_R *satisfies* (PS),

(*iv*) $C^q(\Phi_R, \infty) \approx C^q(\Phi_{(a,b)}, 0) \quad \forall q$.

PROOF. Since $\Phi_{(a,b)}$ has no critical points on $\mathcal{M}$,

$$\delta := \inf_{u \in \mathcal{M}} \left\|\Phi'_{(a,b)}(u)\right\|^* > 0$$

by Lemma 3.3, and then

$$\inf_{u\in\mathcal{M}} \left\|\Phi'_{(a,b)}(Ru)\right\|^* = \delta R^{p-1}, \quad R > 0$$

by homogeneity. It follows that

$$\inf_{u\in\mathcal{M}} \left\|\Phi'(Ru)\right\|^* = (\delta + \mathrm{o}(1))\, R^{p-1} \text{ as } R \to \infty \tag{8.19}$$

since $\sup \|g(R\mathcal{M})\|^* = \mathrm{o}(R^{p-1})$ by (8.17). Take a smooth function $\varphi : [0,\infty) \to [0,1]$ such that $\varphi = 1$ on $[0,1]$ and $\varphi = 0$ on $[2^p, \infty)$ and set

$$\Phi_R(u) = \Phi_{(a,b)}(u) + \varphi(I_p(u)/R^p)\, G(u).$$

Then (i) is clear.

Since

$$\begin{aligned}\left\|(\varphi(I_p/R^p))'(Ru)\right\|^* &= |\varphi'(I_p(Ru)/R^p)|\, \|A_p\,(Ru)\|^*/R^p \\ &= \mathrm{O}(R^{-1}), \quad u \in \mathcal{M}\end{aligned}$$

by (1.5) and $\sup |G(R\mathcal{M})| = \mathrm{o}(R^p)$ by (8.18), (8.19) holds with Φ replaced by Φ_R also. So for sufficiently large R,

$$\inf_{u\in\overline{U_R^c}} \left\|\Phi'(u)\right\|^* > 0, \qquad \inf_{u\in\overline{U_R^c}} \left\|\Phi'_R(u)\right\|^* > 0 \tag{8.20}$$

and hence (ii) follows.

By (8.20), every (PS) sequence for Φ_R has a subsequence in $\overset{\circ}{U}_R$, which then is a (PS) sequence of Φ by (i) and hence has a convergent subsequence.

Since $\Phi_{(a,b)}$ and G are bounded on bounded sets and φ is bounded, Φ_R is also bounded on bounded sets. By (ii), the critical values of Φ_R are bounded from below by $\inf \Phi_R(U_R)$. Taking the a in (3.25) to be less than both $\inf \Phi_R(U_{2R})$ and $\inf \Phi_{(a,b)}(U_{2R})$, say a', gives

$$C^q(\Phi_R, \infty) = H^q(W, \Phi_R^{a'}) = H^q(W, \Phi_{(a,b)}^{a'})$$

since $\Phi_R^{a'}$ and $\Phi_{(a,b)}^{a'}$ lie outside U_{2R}, where $\Phi_R = \Phi_{(a,b)}$ by (i). Since the origin is the only critical point of $\Phi_{(a,b)}$ and $a' < \Phi_{(a,b)}(0)$,

$$H^q(W, \Phi_{(a,b)}^{a'}) \approx C^q(\Phi_{(a,b)}, 0)$$

by Proposition 3.12. □

The main result of this section is

Theorem 8.20. *Assume* (A_1) - (A_4), (B_1) - (B_3), $(B_1^\pm)$ - $(B_4^\pm)$, *and* (8.16) *and* (8.17). *If* (a,b) *is in a path component of* $\mathbb{R}^2\backslash\Sigma(\mathit{\Lambda}_p\,, B_p^+\,, B_p^-\,)$ *containing a point of the form* (λ, λ), *then* (1.1) *has a solution.*

Proof. It suffices to show that the functional Φ_R in Lemma 8.19 has a critical point. Since

$$C^q(\Phi_R, \infty) \approx C^q(\Phi_{(a,b)}, 0) \neq 0$$

for some q by Corollary 8.15, this follows from Proposition 3.16. □

CHAPTER 9

Indefinite Eigenvalue Problems

In this chapter we drop the condition (B_2) in the eigenvalue problem (4.1) and allow J_p to change sign. Then the eigenvalues may be positive or negative, but not zero since $u = 0$ is the only solution of $A_p u = 0$ by (A_2).

9.1. Positive Eigenvalues

First suppose

$(\widetilde{B}_2^+)$ the set

$$\mathcal{M}^+ := \{u \in \mathcal{M} : J_p(u) > 0\}$$

is nonempty.

Since J_p is continuous and even, $\mathcal{M}^+$ is a symmetric open submanifold of $\mathcal{M}$. As in Section 4.1, the positive eigenvalues and the associated eigenvectors on $\mathcal{M}^+$ are the critical values and the critical points of the positive and even C^1-functional

$$\widetilde{\Psi}^+ = \Psi|_{\mathcal{M}^+} = \left.\frac{1}{J_p}\right|_{\mathcal{M}^+}.$$

Since J_p is bounded on the bounded set $\mathcal{M}^+$ by Proposition 1.1, $\widetilde{\Psi}^+$ is bounded from below by a positive constant.

If $(u_j) \subset \mathcal{M}^+$ is a $(\mathrm{PS})_c$ sequence of $\widetilde{\Psi}^+$, then $c \geqslant 0$, and the argument in the proof of Lemma 4.5 shows that $c > 0$ and that a renamed subsequence converges to some $u \in \mathcal{M}$. Since $1/J_p(u_j) \to c$, then $J_p(u) = 1/c > 0$ and hence $u \in \mathcal{M}^+$. So $\widetilde{\Psi}^+$ satisfies (PS). In particular, the set E_λ of eigenvectors of λ that lie on $\mathcal{M}^+$ is compact and the positive spectrum $\sigma^+(A_p, B_p)$ consisting of all positive eigenvalues is closed. We also have the following deformation lemma for $\widetilde{\Psi}^+$ even though $\mathcal{M}^+$ is not complete.

Lemma 9.1. *If $c, \delta > 0$, then there are $\varepsilon > 0$ and $\eta \in C([0,1] \times \mathcal{M}^+, \mathcal{M}^+)$, with $\eta(t, \cdot)$ odd for all $t \in [0,1]$, satisfying*

(*i*) $d(\eta(t,u), u) \leqslant \delta t \quad \forall (t,u) \in [0,1] \times \mathcal{M}^+$,

(*ii*) $\widetilde{\Psi}^+(\eta(t,u)) \leqslant \widetilde{\Psi}^+(u) \quad \forall (t,u) \in [0,1] \times \mathcal{M}^+$,

(*iii*) $\eta(1, (\widetilde{\Psi}^+)^{c+\varepsilon} \backslash N_\delta(K^c)) \subset (\widetilde{\Psi}^+)^{c-\varepsilon}$.

PROOF. We apply Lemma 3.51 to $\Phi = -\left. J_p\right|_{\mathcal{M}}$ at the level $\widetilde{c} = -1/c$. Its critical set at this level is also K^c by Proposition 3.54. An argument similar to that in the proof of Lemma 4.5 shows that Φ satisfies $(\mathrm{PS})_{\widetilde{c}}$. Thus, there are $\widetilde{\varepsilon} > 0$ and $\widetilde{\eta} \in C([0,1] \times \mathcal{M}, \mathcal{M})$, with $\widetilde{\eta}(t, \cdot)$ odd for all $t \in [0,1]$, satisfying

$$d(\widetilde{\eta}(t,u), u) \leqslant \delta t \quad \forall (t,u) \in [0,1] \times \mathcal{M}, \tag{9.1}$$

$$\Phi(\widetilde{\eta}(t,u)) \leqslant \Phi(u) \quad \forall (t,u) \in [0,1] \times \mathcal{M}, \tag{9.2}$$

$$\widetilde{\eta}(1, \Phi^{\widetilde{c}+\widetilde{\varepsilon}} \backslash N_\delta(K^c)) \subset \Phi^{\widetilde{c}-\widetilde{\varepsilon}}. \tag{9.3}$$

If $u \in \mathcal{M}^+$, then

$$J_p(\widetilde{\eta}(t,u)) \geqslant J_p(u) > 0 \tag{9.4}$$

by (9.2), so $\widetilde{\eta}(t,u) \in \mathcal{M}^+$. Let

$$\varepsilon = \frac{c^2 \, \widetilde{\varepsilon}}{1 + c \, \widetilde{\varepsilon}}, \qquad \eta = \widetilde{\eta}|_{[0,1] \times \mathcal{M}^+}.$$

Then (i), (ii), and (iii) follow from (9.1), (9.4), and (9.3), respectively. □

Let $d^+ = i(\mathcal{M}^+)$. Then $1 \leqslant d^+ \leqslant d$ by $(\widetilde{B}_2^+)$, (i_1), (i_2), and (i_8). Let

$$\mathcal{K}^+ = \left\{k \in \mathbb{N} : k \leqslant d^+\right\} \tag{9.5}$$

and for $k \in \mathcal{K}^+$, let

$$\mathcal{F}_k^+ = \left\{M \in \mathcal{F}_k : M \subset \mathcal{M}^+\right\}$$

and

$$\lambda_k^+ = \inf_{M \in \mathcal{F}_k^+} \sup_{u \in M} \widetilde{\Psi}^+(u). \tag{9.6}$$

Since $\mathcal{F}_k^+ \supset \mathcal{F}_{k+1}^+$, $\lambda_k^+ \leqslant \lambda_{k+1}^+$, and since there is a compact set in $\mathcal{F}_k^+$ by Proposition 2.14 (iii), λ_k^+ is finite. When $d^+ < \infty$ we set

$$\lambda_{d^+ + 1}^+ = +\infty. \tag{9.7}$$

The next theorem now follows as in Section 4.2 (Proposition 3.52 applies here since we have Lemma 9.1).

Theorem 9.2. *Assume* (A_1) - (A_4), (B_1), $(\widetilde{B}_2^+)$, *and* (B_3). *Then* $\left(\lambda_k^+\right)_{k \in \mathcal{K}^+}$ *is a nondecreasing sequence of positive eigenvalues of* (4.1).

(i) *If* $\lambda_k^+ = \cdots = \lambda_{k+m-1}^+ = \lambda$, *then* $i(E_\lambda) \geqslant m$. *In particular, there are* d^+ *distinct pairs of eigenvectors on* $\mathcal{M}^+$.

(ii) *The smallest positive eigenvalue is*

$$\lambda_1^+ = \min_{u \in \mathcal{M}^+} \widetilde{\Psi}^+(u) = \min_{J_p(u) > 0} \frac{I_p(u)}{J_p(u)} > 0.$$

(iii) *We have* $i(\mathcal{M}^+ \backslash (\widetilde{\Psi}^+)_{\lambda_k^+}) < k \leqslant i((\widetilde{\Psi}^+)^{\lambda_k^+})$. *If* $\lambda_k^+ < \lambda < \lambda_{k+1}^+$, *then*

$$i((\widetilde{\Psi}^+)^{\lambda_k^+}) = i(\mathcal{M}^+ \backslash (\widetilde{\Psi}^+)_\lambda) = i((\widetilde{\Psi}^+)^\lambda) = i(\mathcal{M}^+ \backslash (\widetilde{\Psi}^+)_{\lambda_{k+1}^+}) = k.$$

(iv) *If* $d^+ = \infty$, *then* $\lambda_k^+ \nearrow +\infty$.

Turning to the critical groups of the functional Φ_λ in (4.10), we have

Theorem 9.3. *Assume* (A_1) - (A_4), (B_1), $(\widetilde{B}_2^+)$, (B_3), *and* $\lambda \notin \sigma^+(A_p\,, B_p)$.

(*i*) *If* $0 \leqslant \lambda < \lambda_1^+$, *then*

$$C^q(\Phi_\lambda, 0) \approx \delta_{q0}\, \mathbb{Z}_2.$$

(*ii*) *If* $\lambda > \lambda_1^+$, *then*

$$C^q(\Phi_\lambda, 0) \approx \widetilde{H}^{q-1}((\widetilde{\Psi}^+)^\lambda) \quad \forall q.$$

In particular, $C^0(\Phi_\lambda, 0) = 0$.

(*iii*) *If* $\lambda_k^+ < \lambda < \lambda_{k+1}^+$, *then*

$$C^k(\Phi_\lambda, 0) \neq 0.$$

PROOF. For $\lambda \geqslant 0$ and $u \in \mathcal{M}$,

$$\Phi_\lambda(u) \begin{cases} \geqslant 1, & u \notin \mathcal{M}^+ \\ = 1 - \dfrac{\lambda}{\widetilde{\Psi}^+(u)}, & u \in \mathcal{M}^+, \end{cases}$$

so $\Phi_\lambda^0 \cap \mathcal{M} = (\widetilde{\Psi}^+)^\lambda$. Since $(\widetilde{\Psi}^+)^\lambda = \varnothing$ if and only if $\lambda < \lambda_1^+$ by Theorem 9.2 (*ii*), (*i*) and (*ii*) follow from (4.13). Since $i((\widetilde{\Psi}^+)^\lambda) = k$ when $\lambda_k^+ < \lambda < \lambda_{k+1}^+$ by Theorem 9.2 (*iii*), then (*iii*) follows from Proposition 2.14 (*iv*). □

9.2. Negative Eigenvalues

Now suppose

$(\widetilde{B}_2^-)$ the set

$$\mathcal{M}^- := \{u \in \mathcal{M} : J_p(u) < 0\}$$

is nonempty.

Applying the preceding discussion to $-B$, we see that $\mathcal{M}^-$ is also a symmetric open submanifold of $\mathcal{M}$ and that the negative eigenvalues and the associated eigenvectors of (4.1) are the critical values and the critical points of the negative and even C^1-functional

$$\widetilde{\Psi}^- = \Psi|_{\mathcal{M}^-} = \left.\frac{1}{J_p}\right|_{\mathcal{M}^-},$$

which is bounded from above by a negative constant and satisfies (PS). The set E_λ of eigenvectors of λ that lie on $\mathcal{M}^-$ is compact and the negative spectrum $\sigma^-(A_p\,, B_p)$ consisting of all negative eigenvalues is closed. The index d^- of $\mathcal{M}^-$ satisfies $1 \leqslant d^- \leqslant d$. Let

$$\mathcal{K}^- = \{k \in \mathbb{N} : k \leqslant d^-\} \tag{9.8}$$

and for $k \in \mathcal{K}^-$, let

$$\mathcal{F}_k^- = \{M \in \mathcal{F}_k : M \subset \mathcal{M}^-\}$$

and

$$\lambda_k^- = \sup_{M \in \mathcal{F}_k^-} \inf_{u \in M} \widetilde{\Psi}^-(u). \tag{9.9}$$

Then λ_k^- is finite and $\lambda_{k+1}^- \leqslant \lambda_k^-$. When $d^- < \infty$ we set

$$\lambda_{d^-+1}^- = -\infty. \tag{9.10}$$

We have

Theorem 9.4. *Assume* (A_1) - (A_4), (B_1), $(\widetilde{B}_2^-)$, *and* (B_3). *Then* $(\lambda_k^-)_{k \in \mathcal{K}^-}$ *is a nonincreasing sequence of negative eigenvalues of* (4.1).

(*i*) *If* $\lambda_k^- = \cdots = \lambda_{k+m-1}^- = \lambda$, *then* $i(E_\lambda) \geqslant m$. *In particular, there are* d^- *distinct pairs of eigenvectors on* $\mathcal{M}^-$.

(*ii*) *The largest negative eigenvalue is*

$$\lambda_1^- = \max_{u \in \mathcal{M}^-} \widetilde{\Psi}^-(u) = \max_{J_p(u)<0} \frac{I_p(u)}{J_p(u)} < 0.$$

(*iii*) *We have* $i(\mathcal{M}^- \backslash (\widetilde{\Psi}^-)^{\lambda_k^-}) < k \leqslant i((\widetilde{\Psi}^-)_{\lambda_k^-})$. *If* $\lambda_{k+1}^- < \lambda < \lambda_k^-$, *then*

$$i((\widetilde{\Psi}^-)_{\lambda_k^-}) = i(\mathcal{M}^- \backslash (\widetilde{\Psi}^-)^{\lambda}) = i((\widetilde{\Psi}^-)_{\lambda}) = i(\mathcal{M}^- \backslash (\widetilde{\Psi}^-)^{\lambda_{k+1}^-}) = k.$$

(*iv*) *If* $d^- = \infty$, *then* $\lambda_k^- \searrow -\infty$.

Theorem 9.5. *Assume* (A_1) - (A_4), (B_1), $(\widetilde{B}_2^-)$, (B_3), *and* $\lambda \notin \sigma^-(A_p, B_p)$.

(*i*) *If* $\lambda_1^- < \lambda \leqslant 0$, *then*

$$C^q(\Phi_\lambda, 0) \approx \delta_{q0}\, \mathbb{Z}_2.$$

(*ii*) *If* $\lambda < \lambda_1^-$, *then*

$$C^q(\Phi_\lambda, 0) \approx \widetilde{H}^{q-1}((\widetilde{\Psi}^-)_\lambda) \quad \forall q.$$

In particular, $C^0(\Phi_\lambda, 0) = 0$.

(*iii*) *If* $\lambda_{k+1}^- < \lambda < \lambda_k^-$, *then*

$$C^k(\Phi_\lambda, 0) \neq 0.$$

9.3. General Case

Since $0 \notin \sigma(A_p, B_p)$,

$$\sigma^+(A_p, B_p) \cup \sigma^-(A_p, B_p) = \sigma(A_p, B_p).$$

When neither, one, or both of $\sigma^\pm(A_p, B_p)$ are empty, combining Theorems 9.3 and 9.5 gives

Theorem 9.6. *Assume* (A_1) - (A_4), (B_1), (B_3), *and* $\lambda \notin \sigma(A_p, B_p)$.

(*i*) *If* $\lambda_1^- < \lambda < \lambda_1^+$, *then*

$$C^q(\Phi_\lambda, 0) \approx \delta_{q0}\, \mathbb{Z}_2.$$

(*ii*) *If* $\lambda < \lambda_1^-$ (resp. $> \lambda_1^+$), *then*

$$C^q(\Phi_\lambda, 0) \approx \widetilde{H}^{q-1}((\widetilde{\Psi}^-)_\lambda) \text{ (resp. } \widetilde{H}^{q-1}((\widetilde{\Psi}^+)^\lambda)) \quad \forall q.$$

In particular, $C^0(\Phi_\lambda, 0) = 0$ *if* $\lambda < \lambda_1^-$ *or* $\lambda > \lambda_1^+$.

(*iii*) *If* $\lambda^-_{k+1} < \lambda < \lambda^-_k$ *or* $\lambda^+_k < \lambda < \lambda^+_{k+1}$, *then*

$$C^k(\Phi_\lambda, 0) \neq 0.$$

In particular, Corollary 4.10 holds without (B_2).

We emphasize that Theorem 9.6 applies in all possible cases:

(*a*) J_p changes sign: $\cdots < \lambda^-_1 < 0 < \lambda^+_1 < \cdots$,
(*b*) $J_p \geqslant 0$ and $\not\equiv 0$: $-\infty = \lambda^-_1 < 0 < \lambda^+_1 < \cdots$,
(*c*) $J_p \leqslant 0$ and $\not\equiv 0$: $\cdots < \lambda^-_1 < 0 < \lambda^+_1 = +\infty$,
(*d*) $J_p \equiv 0$: $-\infty = \lambda^-_1 < \lambda^+_1 = +\infty$.

Example 9.7. When V is allowed to change sign in Example 4.2,

$$J_p(u) = \frac{1}{p} \int_\Omega V(x)\, |u|^p, \qquad d^\pm = \begin{cases} \infty, & |\Omega^\pm| > 0 \\ 0, & |\Omega^\pm| = 0 \end{cases}$$

where $\Omega^\pm = \big\{x \in \Omega : V(x) \gtrless 0\big\}$.

Our setting also includes problems where the eigenvalue appears in the boundary conditions as the next example shows.

Example 9.8. In problem (1.16), taking

$$\big\langle B_p\, u, v\big\rangle = \int_{\partial\Omega} V(x)\, |u|^{p-2}\, uv$$

with $V \in L^s(\partial\Omega)$ and

$$s \begin{cases} > \dfrac{n-1}{p-1}, & p \leqslant n \\ = 1, & p > n \end{cases}$$

gives

$$\begin{cases} -\Delta_p\, u + a(x)\, |u|^{p-2}\, u = 0 & \text{in } \Omega \\ |\nabla u|^{p-2}\, \dfrac{\partial u}{\partial \nu} = \lambda\, V(x)\, |u|^{p-2}\, u & \text{on } \partial\Omega. \end{cases}$$

Then

$$J_p(u) = \frac{1}{p} \int_{\partial\Omega} V(x)\, |u|^p, \qquad d^\pm = \begin{cases} \infty, & |\Gamma^\pm| > 0 \\ 0, & |\Gamma^\pm| = 0 \end{cases}$$

where $\Gamma^\pm = \big\{x \in \partial\Omega : V(x) \gtrless 0\big\}$ and $|\Gamma^\pm|$ is the area of $\Gamma^\pm$.

9.4. Critical Groups of Perturbed Problems

Returning to the equation (1.1), suppose

$$F(u) = \lambda\, J_p(u) + \mathrm{o}(\|u\|^p) \text{ as } u \to 0. \tag{9.11}$$

The following proposition can be combined with the results on the critical groups of Φ at infinity in Chapter 5 to obtain nontrivial solutions.

Proposition 9.9. *Assume* (A_1) - (A_4), (B_1), (B_3), (9.11), *and the origin is an isolated critical point of* Φ.

(*i*) *If* $\lambda_1^- < \lambda < \lambda_1^+$, *then*

$$C^q(\Phi, 0) \approx \delta_{q0}\,\mathbb{Z}_2.$$

(*ii*) *If* $\lambda_{k+1}^- < \lambda < \lambda_k^-$ *or* $\lambda_k^+ < \lambda < \lambda_{k+1}^+$, *then*

$$C^k(\Phi, 0) \neq 0.$$

Proof. (*i*) Since $I_p(u)/J_p(u) \geqslant \lambda_1^+$ (resp. $\leqslant \lambda_1^-$) when $J_p(u) > 0$ (resp. < 0) by Theorem 9.2 (*ii*) (resp. Theorem 9.4 (*ii*)),

$$\frac{\Phi(u)}{I_p(u)} \geqslant \begin{cases} 1 - \dfrac{\lambda}{\lambda_1^+} + \mathrm{o}(1), & \pi_{\mathcal{M}}(u) \in \mathcal{M}^+,\ \lambda > 0 \\ 1 - \dfrac{\lambda}{\lambda_1^-} + \mathrm{o}(1), & \pi_{\mathcal{M}}(u) \in \mathcal{M}^-,\ \lambda < 0 \\ 1 + \mathrm{o}(1), & \text{otherwise} \end{cases} \tag{9.12}$$

as $u \to 0$ by (9.11), where $\pi_{\mathcal{M}}$ is the radial projection onto $\mathcal{M}$, so the origin is a local minimizer of Φ and the conclusion follows from Proposition 3.18.

(*ii*) We show that Φ has a cohomological local splitting near zero in dimension k, with $U = rI\mathcal{M}$ and $h(u) = ru$ where $r > 0$ is sufficiently small, and apply Proposition 3.34.

Case 1: $\lambda_k^+ < \lambda < \lambda_{k+1}^+$. Let $A_0 = (\widetilde{\Psi}^+)^{\lambda_k^+}$ and $B_0 = (\widetilde{\Psi}^+)_{\lambda_{k+1}^+} \cup (\mathcal{M}\backslash\mathcal{M}^+)$. Theorem 9.2 (*iii*) gives (3.32) with $q = k$, and (3.33) follows for sufficiently small r since

$$\frac{\Phi(u)}{I_p(u)} \begin{cases} \leqslant -\left(\dfrac{\lambda}{\lambda_k^+} - 1 + \mathrm{o}(1)\right), & \pi_{\mathcal{M}}(u) \in (\widetilde{\Psi}^+)^{\lambda_k^+} \\ \geqslant 1 - \dfrac{\lambda}{\lambda_{k+1}^+} + \mathrm{o}(1), & \pi_{\mathcal{M}}(u) \in (\widetilde{\Psi}^+)_{\lambda_{k+1}^+} \\ \geqslant 1 + \mathrm{o}(1), & \pi_{\mathcal{M}}(u) \in \mathcal{M}\backslash\mathcal{M}^+ \end{cases} \tag{9.13}$$

as $u \to 0$ by (9.11).

Case 2: $\lambda_{k+1}^- < \lambda < \lambda_k^-$. Let $A_0 = (\widetilde{\Psi}^-)_{\lambda_k^-}$ and $B_0 = (\widetilde{\Psi}^-)^{\lambda_{k+1}^-} \cup (\mathcal{M}\backslash\mathcal{M}^-)$. Theorem 9.4 (*iii*) gives (3.32) with $q = k$, and (3.33) follows for sufficiently small r since

$$\frac{\Phi(u)}{I_p(u)} \begin{cases} \leqslant -\left(\dfrac{\lambda}{\lambda_k^-} - 1 + \mathrm{o}(1)\right), & \pi_{\mathcal{M}}(u) \in (\widetilde{\Psi}^-)_{\lambda_k^-} \\ \geqslant 1 - \dfrac{\lambda}{\lambda_{k+1}^-} + \mathrm{o}(1), & \pi_{\mathcal{M}}(u) \in (\widetilde{\Psi}^-)^{\lambda_{k+1}^-} \\ \geqslant 1 + \mathrm{o}(1), & \pi_{\mathcal{M}}(u) \in \mathcal{M}\backslash\mathcal{M}^- \end{cases} \tag{9.14}$$

as $u \to 0$ by (9.11). □

Now suppose

$$F(u) = \lambda\, J_p(u) + \mathrm{o}(\|u\|^p) \quad \text{as} \quad \|u\| \to \infty. \tag{9.15}$$

We have

Proposition 9.10. *Assume* (A_1) - (A_4), (B_1), (B_3), (9.15), *and* Φ *satisfies* (C).

(*i*) *If* $\lambda_1^- < \lambda < \lambda_1^+$, *then*

$$C^q(\Phi, \infty) \approx \delta_{q0}\, \mathbb{Z}_2.$$

(*ii*) *If* $\lambda_{k+1}^- < \lambda < \lambda_k^-$ *or* $\lambda_k^+ < \lambda < \lambda_{k+1}^+$, *then*

$$C^0(\Phi, \infty) = 0, \quad C^k(\Phi, \infty) \neq 0.$$

Proof. (*i*) (9.12) holds as $\|u\| \to \infty$ also by (9.15) and (1.6), so Φ is bounded from below and the conclusion follows from Corollary 3.19.

(*ii*) We apply Propositions 3.25 and 3.26 with $q = k-1$ and R sufficiently large.

Case 1: $\lambda_k^+ < \lambda < \lambda_{k+1}^+$. Let $A_0 = (\widetilde{\Psi}^+)^{\lambda_k^+}$ and $B_0 = \big[(\widetilde{\Psi}^+)_{\lambda_{k+1}^+} \cup (\mathcal{M} \backslash \mathcal{M}^+)\big]$. Theorem 9.2 (*iii*) gives (3.28). (9.13) holds as $\|u\| \to \infty$ also by (9.15) and (1.6), so Φ is bounded from below on B and (3.31) follows for any $a < \inf \Phi(B)$ and sufficiently large R.

Case 2: $\lambda_{k+1}^- < \lambda < \lambda_k^-$. Let $A_0 = (\widetilde{\Psi}^-)_{\lambda_k^-}$ and $B_0 = \big[(\widetilde{\Psi}^-)^{\lambda_{k+1}^-} \cup (\mathcal{M} \backslash \mathcal{M}^-)\big]$. Theorem 9.4 (*iii*) gives (3.28). (9.14) holds as $\|u\| \to \infty$ also by (9.15) and (1.6), so Φ is bounded from below on B and (3.31) follows for any $a < \inf \Phi(B)$ and sufficiently large R. □

Example 9.11. In Example 9.7, (9.11) (resp. (9.15)) holds if

$$f(x,t) = \lambda\, V(x)\, |t|^{p-2}\, t - g(x,t)$$

for some $g \in \mathrm{Car}(\Omega \times \mathbb{R})$ with subcritical growth and primitive G satisfying

$$G(x,t) = \mathrm{o}(|t|^p) \quad \text{as} \quad |t| \to 0 \text{ (resp. } \infty), \quad \text{uniformly in } x.$$

Here

$$\begin{cases} -\Delta_p\, u = \lambda\, V(x)\, |u|^{p-2}\, u - g(x,u) & \text{in } \Omega \\ \\ u = 0 & \text{on } \partial\Omega. \end{cases}$$

Example 9.12. In Example 9.8, (9.11) (resp. (9.15)) holds if

$$F(x,t) = \mathrm{o}(|t|^p) \quad \text{as} \quad |t| \to 0 \text{ (resp. } \infty), \quad \text{uniformly in } x.$$

Here

$$\begin{cases} -\Delta_p\, u + a(x)\, |u|^{p-2}\, u = f(x,u) & \text{in } \Omega \\ \\ |\nabla u|^{p-2}\, \dfrac{\partial u}{\partial \nu} = \lambda\, V(x)\, |u|^{p-2}\, u & \text{on } \partial\Omega. \end{cases}$$

CHAPTER 10

Anisotropic Systems

In this closing chapter we consider systems of equations of the form (1.1) where each equation may have a different p.

Let $m \in \mathbb{N}$ and for $i = 1, \ldots, m$, let $(W_i, \|\cdot\|_i)$ be a real reflexive Banach space of dimension d_i, $1 \leqslant d_i \leqslant \infty$, with the dual $(W_i^*, \|\cdot\|_i^*)$ and the duality pairing $\langle \cdot, \cdot \rangle_i$. Then their product

$$W = W_1 \times \cdots \times W_m = \{u = (u_1, \ldots, u_m) : u_i \in W_i\}$$

is also a reflexive Banach space with the norm

$$\|u\| = \left(\sum_{i=1}^m \|u_i\|_i^2 \right)^{\frac{1}{2}}$$

and has the dual

$$W^* = W_1^* \times \cdots \times W_m^* = \{L = (L_1, \ldots, L_m) : L_i \in W_i^*\},$$

with the pairing

$$\langle L, u \rangle = \sum_{i=1}^m \langle L_i, u_i \rangle_i$$

and the dual norm

$$\|L\|^* = \left(\sum_{i=1}^m \left(\|L_i\|_i^* \right)^2 \right)^{\frac{1}{2}}.$$

We consider the system of equations

$$A_p\, u = F'(u) \tag{10.1}$$

in W^*, where $p = (p_1, \ldots, p_m)$ with each $p_i \in (1, \infty)$,

$$A_p\, u = (A_{p_1} u_1, \ldots, A_{p_m} u_m),$$

$A_{p_i} \in C(W_i, W_i^*)$ is

(A_{i1}) $(p_i - 1)$-homogeneous and odd,

(A_{i2}) uniformly positive: $\exists\, c_0 > 0$ such that

$$\langle A_{p_i} u_i, u_i \rangle_i \geqslant c_0 \|u_i\|_i^{p_i} \quad \forall u_i \in W_i,$$

(A_{i3}) a potential operator,

A_p satisfies (A_4), and $F \in C^1(W, \mathbb{R})$, $F(0) = 0$, and $F' = (F_{u_1}, \dots, F_{u_m}) : W \to W^*$ is compact.

As in Chapter 1, the potential I_p of A_p satisfying $I_p(0) = 0$ is given by

$$I_p(u) = \sum_{i=1}^m \frac{1}{p_i} \langle A_{p_i} u_i, u_i \rangle_i \,, \tag{10.2}$$

F can be written as

$$F(u) = \sum_{i=1}^m \int_0^1 \langle F_{u_i}(tu), u_i \rangle_i \, dt,$$

and the variational functional associated with equation (10.1) is

$$\begin{aligned} \Phi(u) &= I_p(u) - F(u) \\ &= \sum_{i=1}^m \left[\frac{1}{p_i} \langle A_{p_i} u_i, u_i \rangle_i - \int_0^1 \langle F_{u_i}(tu), u_i \rangle_i \, dt \right], \quad u \in W. \end{aligned}$$

Although not homogeneous, I_p is even and satisfies

$$c_0 \sum_{i=1}^m \frac{1}{p_i} \|u_i\|_i^{p_i} \leqslant I_p(u) \leqslant C_0 \sum_{i=1}^m \frac{1}{p_i} \|u_i\|_i^{p_i} \quad \forall u \in W \tag{10.3}$$

for some $C_0 \geqslant c_0$. Writing $(u_1/p_1, \dots, u_m/p_m) = u/p$, we also have

$$\langle I_p'(u), u/p \rangle = \langle A_p \, u, u/p \rangle = I_p(u), \tag{10.4}$$

analogous to (4.3).

Proposition 10.1. *If each W_i is uniformly convex and*

$$\langle A_{p_i} u_i, v_i \rangle_i \leqslant r_i \|u_i\|_i^{p_i - 1} \|v_i\|_i \,, \quad \langle A_{p_i} u_i, u_i \rangle_i = r_i \|u_i\|_i^{p_i} \quad \forall u_i, v_i \in W_i$$

for some $r_i > 0$, then A_p satisfies (A_4).

Proof. If $u^j \rightharpoonup u$ and $\langle A_p \, u^j, u^j - u \rangle \to 0$, then

$$\begin{aligned} 0 &\leqslant \sum_{i=1}^m r_i \left(\left\| u_i^j \right\|_i^{p_i - 1} - \|u_i\|_i^{p_i - 1} \right) \left(\left\| u_i^j \right\|_i - \|u_i\|_i \right) \\ &\leqslant \sum_{i=1}^m \left(\left\langle A_{p_i} u_i^j, u_i^j \right\rangle_i - \left\langle A_{p_i} u_i^j, u_i \right\rangle_i - \left\langle A_{p_i} u_i, u_i^j \right\rangle_i + \langle A_{p_i} u_i, u_i \rangle_i \right) \\ &= \left\langle A_p \, u^j, u^j - u \right\rangle - \left\langle A_p \, u, u^j - u \right\rangle \to 0, \end{aligned}$$

so, for each i, $\left\| u_i^j \right\|_i \to \|u_i\|_i$ and hence $u_i^j \to u_i$ by uniform convexity. □

Example 10.2. Consider the p-Laplacian system

$$\begin{cases} -\Delta_p \, u = \nabla_u \, F(x, u) & \text{in } \Omega \\ u = 0 & \text{on } \partial\Omega \end{cases} \tag{10.5}$$

where $p = (p_1, \dots, p_m)$ with each $p_i \in (1, \infty)$, $u = (u_1, \dots, u_m)$, $\Delta_p u = (\Delta_{p_1} u_1, \dots, \Delta_{p_m} u_m)$, and $F \in C^1(\Omega \times \mathbb{R}^m)$ satisfies $F(x, 0) \equiv 0$ and

$$\left| \frac{\partial F}{\partial u_i} \right| \leqslant C \left(\sum_{j=1}^{m} |u_j|^{r_{ij}-1} + 1 \right) \quad \forall (x, u) \in \Omega \times \mathbb{R}^m \tag{10.6}$$

for some $r_{ij} \in (1, 1 + p_j^*(p_i^* - 1)/p_i^*)$. Problem (1.21) is the special case $p_1 = \dots = p_m = p$. Here $W_i = W_0^{1,p_i}(\Omega)$,

$$\langle A_{p_i} u_i, v_i \rangle_i = \int_\Omega |\nabla u_i|^{p_i-2} \nabla u_i \cdot \nabla v_i, \qquad I_p(u) = \sum_{i=1}^{m} \frac{1}{p_i} \int_\Omega |\nabla u_i|^{p_i},$$

and the operator F is still given by (1.22). A_p satisfies (A_4) by Proposition 10.1.

10.1. Eigenvalue Problems

We recall that a (continuous) representation of a topological group G on a Banach space W is a homomorphism $\rho : G \to GL(W)$, where $GL(W)$ is the group of invertible bounded linear operators on W, such that the map

$$G \times W \to W, \quad (\alpha, u) \mapsto \rho(\alpha)\, u$$

is continuous. Then G also has the dual representation

$$\rho^* : G \to GL(W^*), \quad \langle \rho^*(\alpha)\, L, u \rangle = \big\langle L, \rho(\alpha^{-1})\, u \big\rangle$$

on the dual space W^*.

For $s = (s_1, \dots, s_m)$ with each $s_i \in (0, \infty)$, the multiplicative group $G = \mathbb{R} \backslash \{0\}$ of all nonzero real numbers has the representation

$$\rho_s(\alpha)\, u = (|\alpha|^{s_1-1} \alpha\, u_1, \dots, |\alpha|^{s_m-1} \alpha\, u_m)$$

on $W = W_1 \times \dots \times W_m$, with the dual representation

$$\rho_s^*(\alpha)\, L = (|\alpha|^{-s_1-1} \alpha\, L_1, \dots, |\alpha|^{-s_m-1} \alpha\, L_m)$$

on $W^* = W_1^* \times \dots \times W_m^*$.

We write $(1/p_1, \dots, 1/p_m) = 1/p$ and $(1 - 1/p_1, \dots, 1 - 1/p_m) = 1 - 1/p$, and let

$$u_\alpha = \rho_{1/p}(\alpha)\, u = (|\alpha|^{1/p_1-1} \alpha\, u_1, \dots, |\alpha|^{1/p_m-1} \alpha\, u_m)$$

and

$$L^\alpha = \rho_{1-1/p}^*(\alpha)\, L = (|\alpha|^{1/p_1-2} \alpha\, L_1, \dots, |\alpha|^{1/p_m-2} \alpha\, L_m).$$

Then for any $\Phi \in C^1(W, \mathbb{R})$,

$$\begin{aligned} (\Phi(u_\alpha))' &= (|\alpha|^{1/p_1-1} \alpha\, \Phi_{u_1}(u_\alpha), \dots, |\alpha|^{1/p_m-1} \alpha\, \Phi_{u_m}(u_\alpha)) \\ &= |\alpha|\, \Phi'(u_\alpha)^\alpha \end{aligned} \tag{10.7}$$

and

$$\frac{d}{d\alpha}\left(\Phi(u_\alpha)\right) = \sum_{i=1}^m \left\langle \Phi_{u_i}(u_\alpha), |\alpha|^{1/p_i-2}\,\alpha\, u_i/p_i \right\rangle_i \tag{10.8}$$

$$= |\alpha|^{-1}\left\langle \Phi'(u_\alpha), u_\alpha/p \right\rangle = \left\langle \Phi'(u_\alpha)^\alpha, u/p \right\rangle.$$

We have

$$(A_p\, u_\alpha)^\alpha = \left(|\alpha|^{-1/p_1}\,\alpha\, A_{p_1} u_1, \dots, |\alpha|^{-1/p_m}\,\alpha\, A_{p_m} u_m\right)^\alpha = A_p\, u \tag{10.9}$$

and

$$I_p(u_\alpha) = \sum_{i=1}^m \frac{1}{p_i}\left\langle |\alpha|^{-1/p_i}\,\alpha\, A_{p_i} u_i, |\alpha|^{1/p_i-1}\,\alpha\, u_i \right\rangle_i = |\alpha|\, I_p(u), \tag{10.10}$$

so we make

Definition 10.3. Let $\mathcal{J}$ be the subspace of $C^1(W, \mathbb{R})$ consisting of all functionals J satisfying

$$J(u_\alpha) = |\alpha|\, J(u) \quad \forall u \in W,\ \alpha \in G. \tag{10.11}$$

Then we have

Proposition 10.4. *If $J \in \mathcal{J}$, then*

(*i*) $J(0) = 0$,
(*ii*) *J is even,*
(*iii*) $J'(u_\alpha)^\alpha = J'(u) \quad \forall u \in W,\ \alpha \in G$,
(*iv*) $\langle J'(u), u/p\rangle = J(u) \quad \forall u \in W$.

PROOF. (*i*) Take the limit of (10.11) as $\alpha \to 0$ and note that $u_\alpha \to 0$.
(*ii*) Take $\alpha = -1$ in (10.11) and note that $u_{-1} = -u$.
(*iii*) Differentiate (10.11) with respect to u using (10.7).
(*iv*) Differentiate (10.11) with respect to α at $\alpha = 1$ using (10.8). □

Now we consider the eigenvalue problem

$$A_p\, u = \lambda\, J'(u) \tag{10.12}$$

in W^*, where $J \in \mathcal{J}$ with J' compact. (10.9) and Proposition 10.4 (*iii*) imply that if u is an eigenvector associated with λ, then so is u_α for any $\alpha \in G$ and hence the entire orbit

$$\mathcal{O}(u) = \left\{u_\alpha : \alpha \in G\right\}$$

of u is in the solution set of (10.12). We set $u_0 = 0$ for convenience, and note that the map $\mathbb{R} \times W \to W$, $(\alpha, u) \mapsto u_\alpha$ is continuous.

Example 10.5. In problem (10.5), taking

$$\langle A_{p_i} u_i, v_i\rangle_i = r_i \int_\Omega |\nabla u_i|^{p_i-2}\, \nabla u_i \cdot \nabla v_i, \qquad I_p(u) = \sum_{i=1}^m \frac{r_i}{p_i} \int_\Omega |\nabla u_i|^{p_i}$$

and

$$J(u) = \int_\Omega V(x)\, |u_1|^{r_1} \cdots |u_m|^{r_m}$$

with $r_i \in (1, p_i)$,

$$\sum_{i=1}^{m} \frac{r_i}{p_i} = 1, \tag{10.13}$$

and $V \in L^\infty(\Omega)$ gives

$$\begin{cases} -\Delta_{p_i} u_i = \lambda\, V(x)\, |u_1|^{r_1} \cdots |u_i|^{r_i - 2} \cdots |u_m|^{r_m} u_i & \text{in } \Omega,\ i = 1, \dots, m \\ \\ \quad u_1 = \cdots = u_m = 0 & \text{on } \partial\Omega. \end{cases}$$

A_p satisfies (A_4) by Proposition 10.1 again, and (10.13) implies (10.11).

By (10.4), zero is the only critical value of I_p and hence it follows from the implicit function theorem that

$$\mathcal{M} := \big\{u \in W : I_p(u) = 1\big\}$$

is a C^1-Finsler manifold. Moreover, $\mathcal{M}$ is complete and symmetric since I_p is continuous and even. By (10.3),

$$0 < \inf_{u \in \mathcal{M}} \|u\| \leqslant \sup_{u \in \mathcal{M}} \|u\| < \infty. \tag{10.14}$$

The following lemma implies that $\mathcal{M}$ is radially homeomorphic to the unit sphere S in W.

Lemma 10.6. *There is a C^1-map $T : S \to (0, \infty)$ such that*

$$\mathcal{M} = \big\{T(u)\, u : u \in S\big\}.$$

PROOF. We have $I_p(0) = 0$, and for $u \in S$ and $t \geqslant 0$,

$$I_p(tu) = \sum_{i=1}^{m} \frac{t^{p_i}}{p_i} \left\langle A_{p_i} u_i, u_i \right\rangle_i \geqslant c_0 \sum_{i=1}^{m} \frac{t^{p_i}}{p_i} \|u_i\|_i^{p_i} \to \infty \text{ as } t \to \infty$$

by (10.2), (A_{i1}), and (A_{i2}), and

$$\frac{d}{dt}\big(I_p(tu)\big) = \sum_{i=1}^{m} t^{p_i - 1} \left\langle A_{p_i} u_i, u_i \right\rangle_i \geqslant c_0 \sum_{i=1}^{m} t^{p_i - 1} \|u_i\|_i^{p_i} > 0, \quad t > 0.$$

So there is a unique $T(u) > 0$ such that

$$t < \text{ (resp. } =, >) \ T(u) \implies I_p(tu) < \text{ (resp. } =, >)\ 1$$

and the map T is C^1 by the implicit function theorem. □

As in Chapter 9, we first consider the case where the set

$$\mathcal{M}^+ := \big\{u \in \mathcal{M} : J(u) > 0\big\}$$

is nonempty. Since J is continuous and even, $\mathcal{M}^+$ is a symmetric open submanifold of $\mathcal{M}$. We will show that the positive eigenvalues and the associated eigenvectors on $\mathcal{M}^+$ are the critical values and the critical points of the positive and even C^1-functional

$$\widetilde{\Psi}^+ = \frac{1}{J}\bigg|_{\mathcal{M}^+}.$$

Since J is bounded on the bounded set $\mathcal{M}^+$, $\widetilde{\Psi}^+$ is bounded from below by a positive constant.

Since

$$\left(\frac{1}{J}\right)'(u) = -\frac{J'(u)}{J(u)^2} = -\widetilde{\Psi}^+(u)^2\, J'(u), \qquad I_p'(u) = A_p\, u,$$

it follows from Proposition 3.54 that the norm of $(\widetilde{\Psi}^+)'(u) \in T_u^*\mathcal{M}$ is given by

$$\left\|(\widetilde{\Psi}^+)'(u)\right\|_u^* = \min_{\mu \in \mathbb{R}} \left\|\mu\, A_p\, u + \widetilde{\Psi}^+(u)^2\, J'(u)\right\|^*. \tag{10.15}$$

Lemma 10.7. *Positive eigenvalues of* (10.12) *coincide with the critical values of* $\widetilde{\Psi}^+$, *i.e.,* $\lambda > 0$ *is an eigenvalue if and only if there is a* $u \in \mathcal{M}^+$ *such that* $(\widetilde{\Psi}^+)'(u) = 0$ *and* $\widetilde{\Psi}^+(u) = \lambda$.

PROOF. By (10.15), $(\widetilde{\Psi}^+)'(u) = 0$ if and only if

$$\mu\, A_p\, u + \widetilde{\Psi}^+(u)^2\, J'(u) = 0 \tag{10.16}$$

for some $\mu \in \mathbb{R}$. If (10.16) holds, applying it to u/p and using (10.4) and Proposition 10.4 (iv) gives

$$\mu = -\widetilde{\Psi}^+(u)^2\, \frac{\langle J'(u), u/p\rangle}{\langle A_p\, u, u/p\rangle} = -\widetilde{\Psi}^+(u)^2\, \frac{J(u)}{I_p(u)} = -\widetilde{\Psi}^+(u) < 0,$$

so (10.16) reduces to (10.12) with $\lambda = \widetilde{\Psi}^+(u)$. Conversely, if $\lambda > 0$ is an eigenvalue and $u \in \mathcal{M}^+$ is an associated eigenvector,

$$\lambda = \frac{\langle A_p\, u, u/p\rangle}{\langle J'(u), u/p\rangle} = \frac{I_p(u)}{J(u)} = \widetilde{\Psi}^+(u),$$

so (10.12) implies (10.16) with $\mu = -\widetilde{\Psi}^+(u)$. □

Example 10.8. In Example 10.5,

$$\mathcal{M} = \left\{(u_1, \dots, u_m) \in W_0^{1,p_1}(\Omega) \times \cdots \times W_0^{1,p_m}(\Omega) : \sum_{i=1}^m \frac{r_i}{p_i} \int_\Omega |\nabla u_i|^{p_i} = 1\right\},$$

$$\widetilde{\Psi}^+(u) = \frac{1}{\displaystyle\int_\Omega V(x)\, |u_1|^{r_1} \cdots |u_m|^{r_m}}.$$

Lemma 10.9. $\widetilde{\Psi}^+$ *satisfies* (PS).

PROOF. Let $(u^j) \subset \mathcal{M}^+$ be a $(\mathrm{PS})_c$ sequence. Then $c \geqslant 0$. By (10.14), (u^j) is bounded, so a renamed subsequence converges weakly to some $u \in W$ since W is reflexive, and $J'(u^j)$ converges to some $L \in W^*$ for a further subsequence since J' is compact. Since the sequence (u_j/p) is also bounded and converges weakly to u/p, then

$$\widetilde{\Psi}^+(u^j) = \frac{1}{\langle J'(u^j), u^j/p\rangle} \to \frac{1}{\langle L, u/p\rangle} \neq 0$$

by Proposition 10.4 (iv) and Lemma 3.4, so $c > 0$.

By (10.15),

$$\mu^j\, A_p\, u^j + \widetilde{\Psi}^+(u^j)^2\, J'(u^j) \to 0 \tag{10.17}$$

for some sequence $(\mu^j) \subset \mathbb{R}$. Applying (10.17) to u^j/p and using (10.4) and Proposition 10.4 (iv) gives $\mu^j + \widetilde{\Psi}^+(u^j) \to 0$, so $\mu^j \to -c \neq 0$. Now applying (10.17) to $u^j - u$ gives $\langle A_p\, u^j, u^j - u\rangle \to 0$ since $\langle J'(u^j), u^j - u\rangle \to 0$ by Lemma 3.4, so $u^j \to u \in \mathcal{M}$ for a subsequence by (A_4). Since $1/J(u_j) \to c$, then $J(u) = 1/c > 0$ and hence $u \in \mathcal{M}^+$. □

Lemma 10.9 implies that the set E_λ of eigenvectors of λ that lie on $\mathcal{M}^+$ is compact and the positive spectrum $\sigma^+(A_p\,, J')$ consisting of all positive eigenvalues is closed.

Applying Lemma 3.51 to $\Phi = -\, J|_{\mathcal{M}}$ at the level $\widetilde{c} = -1/c$, and noting that an argument similar to that in the proof of Lemma 10.9 shows that Φ satisfies $(\mathrm{PS})_{\widetilde{c}}$, we see that Lemma 9.1 still holds. So, defining $\mathcal{K}^+$ and λ_k^+ by (9.5), (9.6), and (9.7), we have the following theorem (the second equality in (10.18) follows from (10.10) and (10.11)).

Theorem 10.10. *Assume (A_{i1}) - (A_{i3}), (A_4), and $J \in \mathcal{J}$ with J' compact. Then $(\lambda_k^+)_{k\in\mathcal{K}^+}$ is a nondecreasing sequence of positive eigenvalues of* (10.12).

(i) If $\lambda_k^+ = \cdots = \lambda_{k+m-1}^+ = \lambda$, then $i(E_\lambda) \geqslant m$. In particular, there are d^+ distinct pairs of eigenvectors on $\mathcal{M}^+$.

(ii) The smallest positive eigenvalue is

$$\lambda_1^+ = \min_{u\in\mathcal{M}^+} \widetilde{\Psi}^+(u) = \min_{\substack{u\in\mathcal{M}^+\\ \alpha\neq 0}} \frac{I_p(u_\alpha)}{J(u_\alpha)} > 0. \tag{10.18}$$

(iii) We have $i(\mathcal{M}^+\backslash(\widetilde{\Psi}^+)_{\lambda_k^+}) < k \leqslant i((\widetilde{\Psi}^+)^{\lambda_k^+})$. If $\lambda_k^+ < \lambda < \lambda_{k+1}^+$, then

$$i((\widetilde{\Psi}^+)^{\lambda_k^+}) = i(\mathcal{M}^+\backslash(\widetilde{\Psi}^+)_\lambda) = i((\widetilde{\Psi}^+)^\lambda) = i(\mathcal{M}^+\backslash(\widetilde{\Psi}^+)_{\lambda_{k+1}^+}) = k.$$

(iv) If $d^+ = \infty$, then $\lambda_k^+ \nearrow +\infty$.

We now consider the case where the set

$$\mathcal{M}^- := \{u \in \mathcal{M} : J(u) < 0\}$$

is nonempty. Applying the preceding discussion to $-J$ as in Chapter 9, we see that $\mathcal{M}^-$ is also a symmetric open submanifold of $\mathcal{M}$ and that the negative eigenvalues and the associated eigenvectors of (10.12) are the critical values and the critical points of the negative and even C^1-functional

$$\widetilde{\Psi}^- = \left.\frac{1}{J}\right|_{\mathcal{M}^-},$$

which is bounded from above by a negative constant and satisfies (PS). The set E_λ of eigenvectors of λ that lie on $\mathcal{M}^-$ is compact and the negative spectrum $\sigma^-(A_p\,, J')$ consisting of all negative eigenvalues is closed. Defining $\mathcal{K}^-$ and λ_k^- by (9.8), (9.9), and (9.10), we have

Theorem 10.11. *Assume (A_{i1}) - (A_{i3}), (A_4), and $J \in \mathcal{J}$ with J' compact. Then $\left(\lambda_k^-\right)_{k\in\mathcal{K}^-}$ is a nonincreasing sequence of negative eigenvalues of (10.12).*

(i) *If $\lambda_k^- = \cdots = \lambda_{k+m-1}^- = \lambda$, then $i(E_\lambda) \geqslant m$. In particular, there are d^- distinct pairs of eigenvectors on $\mathcal{M}^-$.*

(ii) *The largest negative eigenvalue is*

$$\lambda_1^- = \max_{u\in\mathcal{M}^-} \widetilde{\Psi}^-(u) = \max_{\substack{u\in\mathcal{M}^- \\ \alpha\neq 0}} \frac{I_p(u_\alpha)}{J(u_\alpha)} < 0.$$

(iii) *We have $i(\mathcal{M}^-\backslash(\widetilde{\Psi}^-)^{\lambda_k^-}) < k \leqslant i((\widetilde{\Psi}^-)_{\lambda_k^-})$. If $\lambda_{k+1}^- < \lambda < \lambda_k^-$, then*

$$i((\widetilde{\Psi}^-)_{\lambda_k^-}) = i(\mathcal{M}^-\backslash(\widetilde{\Psi}^-)^\lambda) = i((\widetilde{\Psi}^-)_\lambda) = i(\mathcal{M}^-\backslash(\widetilde{\Psi}^-)^{\lambda_{k+1}^-}) = k.$$

(iv) *If $d^- = \infty$, then $\lambda_k^- \searrow -\infty$.*

Since $u = 0$ is the only solution of $A_p\, u = 0$ by (A_{i2}), $0 \notin \sigma(A_p\,, J')$ and hence

$$\sigma^+(A_p\,, J') \cup \sigma^-(A_p\,, J') = \sigma(A_p\,, J').$$

When $\lambda \notin \sigma(A_p\,, J')$, the critical groups of the associated functional

$$\Phi_\lambda(u) = I_p(u) - \lambda\, J(u), \quad u \in W$$

at zero are given by (4.11).

By (10.10) and (10.11),

$$\Phi_\lambda(u_\alpha) = |\alpha|\, \Phi_\lambda(u),$$

so $\mathcal{O}(\Phi_\lambda^0) \subset \Phi_\lambda^0$. Hence Φ_λ^0 contracts to $\{0\}$ via

$$\Phi_\lambda^0 \times [0,1] \to \Phi_\lambda^0, \quad (u,t) \mapsto u_{1-t}.$$

If $u \neq 0$, $\alpha_u := I_p(u) > 0$ by (10.3) and the path component of $\mathcal{O}(u)$ containing u intersects $\mathcal{M}$ at $\widetilde{u} := u_{\alpha_u^{-1}}$ since $I_p(\widetilde{u}) = 1$ by (10.10). Hence $\Phi_\lambda^0\backslash\{0\}$ deformation retracts to $\Phi_\lambda^0 \cap \mathcal{M}$ via

$$(\Phi_\lambda^0\backslash\{0\}) \times [0,1] \to \Phi_\lambda^0\backslash\{0\}, \quad (u,t) \mapsto u_{1-t+t\,\alpha_u^{-1}}.$$

Thus, (4.13) still holds, and as in Chapter 9 we get

Theorem 10.12. *Assume* (A_{i1}) - (A_{i3}), (A_4), $J \in \mathcal{J}$ *with* J' *compact, and* $\lambda \notin \sigma(A_p, J')$.

(*i*) *If* $\lambda_1^- < \lambda < \lambda_1^+$, *then*

$$C^q(\Phi_\lambda, 0) \approx \delta_{q0}\,\mathbb{Z}_2.$$

(*ii*) *If* $\lambda < \lambda_1^-$ (resp. $> \lambda_1^+$), *then*

$$C^q(\Phi_\lambda, 0) \approx \widetilde{H}^{q-1}((\widetilde{\Psi}^-)_\lambda) \text{ (resp. } \widetilde{H}^{q-1}((\widetilde{\Psi}^+)^\lambda)) \quad \forall q.$$

In particular, $C^0(\Phi_\lambda, 0) = 0$ *if* $\lambda < \lambda_1^-$ *or* $\lambda > \lambda_1^+$.

(*iii*) *If* $\lambda_{k+1}^- < \lambda < \lambda_k^-$ *or* $\lambda_k^+ < \lambda < \lambda_{k+1}^+$, *then*

$$C^k(\Phi_\lambda, 0) \neq 0.$$

In particular, $C^q(\Phi_\lambda, 0) \neq 0$ *for some* q.

10.2. Critical Groups of Perturbed Systems

Returning to the system (10.1), suppose

$$F = \lambda\, J - G \tag{10.19}$$

with

$$G(u_\alpha) = \mathrm{o}(\alpha) \quad \text{as } \alpha \searrow 0, \quad \text{uniformly in } u \in \mathcal{M}. \tag{10.20}$$

Proposition 10.13. *Assume* (A_{i1}) - (A_{i3}), (A_4), $J \in \mathcal{J}$ *with* J' *compact,* (10.19) *and* (10.20) *hold, and the origin is an isolated critical point of* Φ.

(*i*) *If* $\lambda_1^- < \lambda < \lambda_1^+$, *then the origin is a local minimizer of* Φ *and*

$$C^q(\Phi, 0) \approx \delta_{q0}\,\mathbb{Z}_2. \tag{10.21}$$

(*ii*) *If* $\lambda_{k+1}^- < \lambda < \lambda_k^-$ *or* $\lambda_k^+ < \lambda < \lambda_{k+1}^+$, *then*

$$C^k(\Phi, 0) \neq 0.$$

PROOF. For $r > 0$,

$$h_r(tu) = u_{rt}, \quad u \in \mathcal{M},\ t \in [0, 1]$$

defines a homeomorphism from $I\mathcal{M}$ onto

$$U_r = \big\{u_\alpha : u \in \mathcal{M},\ 0 \leqslant \alpha \leqslant r\big\}$$

such that $h_r(0) = 0$, with the inverse

$$h_r^{-1}(u) = \begin{cases} \dfrac{\alpha_u}{r}\, \widetilde{u}, & u \neq 0 \\ 0, & u = 0 \end{cases} \tag{10.22}$$

where α_u and $\widetilde{u}$ are as in the last section. In particular, U_r is a neighborhood of zero. By (10.19), (10.20), (10.10), and (10.11),

$$\Phi(u_\alpha) = \alpha\,(1 - \lambda\, J(u) + \mathrm{o}(1)) \quad \text{as } \alpha \searrow 0, \quad \text{uniformly in } u \in \mathcal{M}. \tag{10.23}$$

(*i*) Since $J(u) \leqslant 1/\lambda_1^+$ (resp. $\geqslant 1/\lambda_1^-$) for $u \in \mathcal{M}^+$ (resp. $\mathcal{M}^-$) by Theorem 10.10 (*ii*) (resp. Theorem 10.11 (*ii*)),

$$\Phi(u_\alpha) \geqslant \begin{cases} \alpha\left(1 - \dfrac{\lambda}{\lambda_1^+} + \mathrm{o}(1)\right), & u \in \mathcal{M}^+,\ \lambda > 0 \\ \alpha\left(1 - \dfrac{\lambda}{\lambda_1^-} + \mathrm{o}(1)\right), & u \in \mathcal{M}^-,\ \lambda < 0 \\ \alpha\,(1 + \mathrm{o}(1)), & \text{otherwise} \end{cases} \tag{10.24}$$

by (10.23), and (10.21) follows from Proposition 3.18.

(*ii*) We show that Φ has a cohomological local splitting near zero in dimension k, with $U = U_r$ and $h = h_r$ where $r > 0$ is sufficiently small, and apply Proposition 3.34. Note that

$$A = h_r(IA_0) = \left\{u_\alpha : u \in A_0,\ 0 \leqslant \alpha \leqslant r\right\},$$

$$B = h_r(IB_0) \cup \{0\} = \left\{u_\alpha : u \in B_0,\ 0 < \alpha \leqslant r\right\} \cup \{0\}.$$

Case 1: $\lambda_k^+ < \lambda < \lambda_{k+1}^+$. Let $A_0 = (\widetilde{\Psi}^+)^{\lambda_k^+}$ and $B_0 = (\widetilde{\Psi}^+)_{\lambda_{k+1}^+} \cup (\mathcal{M}\backslash\mathcal{M}^+)$. Theorem 10.10 (*iii*) gives (3.32) with $q = k$, and (3.33) follows for sufficiently small r since

$$\Phi(u_\alpha) \begin{cases} \leqslant -\alpha\left(\dfrac{\lambda}{\lambda_k^+} - 1 + \mathrm{o}(1)\right), & u \in (\widetilde{\Psi}^+)^{\lambda_k^+} \\ \geqslant \alpha\left(1 - \dfrac{\lambda}{\lambda_{k+1}^+} + \mathrm{o}(1)\right), & u \in (\widetilde{\Psi}^+)_{\lambda_{k+1}^+} \\ \geqslant \alpha\,(1 + \mathrm{o}(1)), & u \in \mathcal{M}\backslash\mathcal{M}^+ \end{cases} \tag{10.25}$$

by (10.23).

Case 2: $\lambda_{k+1}^- < \lambda < \lambda_k^-$. Let $A_0 = (\widetilde{\Psi}^-)_{\lambda_k^-}$ and $B_0 = (\widetilde{\Psi}^-)^{\lambda_{k+1}^-} \cup (\mathcal{M}\backslash\mathcal{M}^-)$. Theorem 10.11 (*iii*) gives (3.32) with $q = k$, and (3.33) follows for sufficiently small r since

$$\Phi(u_\alpha) \begin{cases} \leqslant -\alpha\left(\dfrac{\lambda}{\lambda_k^-} - 1 + \mathrm{o}(1)\right), & u \in (\widetilde{\Psi}^-)_{\lambda_k^-} \\ \geqslant \alpha\left(1 - \dfrac{\lambda}{\lambda_{k+1}^-} + \mathrm{o}(1)\right), & u \in (\widetilde{\Psi}^-)^{\lambda_{k+1}^-} \\ \geqslant \alpha\,(1 + \mathrm{o}(1)), & u \in \mathcal{M}\backslash\mathcal{M}^- \end{cases} \tag{10.26}$$

by (10.23). □

Now suppose (10.19) holds with

$$G(u_\alpha) = \mathrm{o}(\alpha) \text{ as } \alpha \nearrow \infty, \text{ uniformly in } u \in \mathcal{M}. \tag{10.27}$$

Proposition 10.14. *Assume* (A_{i1}) - (A_{i3}), (A_4), $J \in \mathcal{J}$ *with* J' *compact,* (10.19) *and* (10.27) *hold, and* Φ *satisfies* (C).

(i) If $\lambda_1^- < \lambda < \lambda_1^+$, then Φ is bounded from below and

$$C^q(\Phi, \infty) \approx \delta_{q0}\, \mathbb{Z}_2. \tag{10.28}$$

(ii) If $\lambda_{k+1}^- < \lambda < \lambda_k^-$ or $\lambda_k^+ < \lambda < \lambda_{k+1}^+$, then

$$C^0(\Phi, \infty) = 0, \quad C^k(\Phi, \infty) \neq 0.$$

PROOF. Identifying W with $\{tu : u \in \mathcal{M},\, t \in [0, \infty)\}$,

$$h_1(tu) = u_t$$

defines a homeomorphism of W (h_1^{-1} is given by (10.22) with $r = 1$). In particular,

$$W = \{u_\alpha : u \in \mathcal{M},\, \alpha \geqslant 0\}.$$

By (10.19), (10.27), (10.10), and (10.11),

$$\Phi(u_\alpha) = \alpha\,(1 - \lambda\, J(u) + \mathrm{o}(1)) \quad \text{as } \alpha \nearrow \infty, \quad \text{uniformly in } u \in \mathcal{M}. \tag{10.29}$$

(i) (10.24) holds as $\alpha \nearrow \infty$ also by (10.29), and (10.28) follows from Corollary 3.19.

(ii) We apply Propositions 3.25 and 3.27 with $q = k - 1$, $h = h_1$, and R sufficiently large. Note that

$$A = h_1(RA_0) = \{u_R : u \in A_0\},$$

$$B = h_1(\pi_{\mathcal{M}}^{-1}(B_0) \cup \{0\}) = \{u_\alpha : u \in B_0,\, \alpha > 0\} \cup \{0\}.$$

Case 1: $\lambda_k^+ < \lambda < \lambda_{k+1}^+$. Let $A_0 = (\widetilde{\Psi}^+)^{\lambda_k^+}$ and $B_0 = \big[(\widetilde{\Psi}^+)_{\lambda_{k+1}^+} \cup (\mathcal{M} \backslash \mathcal{M}^+)\big]$. Theorem 10.10 *(iii)* gives (3.28). (10.25) holds as $\alpha \nearrow \infty$ also by (10.29), so Φ is bounded from below on B and (3.31) follows for any $a < \inf \Phi(B)$ and sufficiently large R.

Case 2: $\lambda_{k+1}^- < \lambda < \lambda_k^-$. Let $A_0 = (\widetilde{\Psi}^-)_{\lambda_k^-}$ and $B_0 = \big[(\widetilde{\Psi}^-)^{\lambda_{k+1}^-} \cup (\mathcal{M} \backslash \mathcal{M}^-)\big]$. Theorem 10.11 *(iii)* gives (3.28). (10.26) holds as $\alpha \nearrow \infty$ also by (10.29), so Φ is bounded from below on B and (3.31) follows for any $a < \inf \Phi(B)$ and sufficiently large R. □

Example 10.15. In Example 10.5, setting

$$G(x, u) = \lambda\, V(x)\, |u_1|^{r_1} \cdots |u_m|^{r_m} - F(x, u)$$

gives

$$\begin{cases} -\Delta_{p_i} u_i = \lambda\, V(x)\, |u_1|^{r_1} \cdots |u_i|^{r_i - 2} \cdots |u_m|^{r_m}\, u_i - \dfrac{\partial G}{\partial u_i}(x, u) & \\ \qquad\qquad\qquad\qquad\qquad\qquad\qquad\qquad\qquad \text{in } \Omega,\ i = 1, \dots, m \\ u_1 = \cdots = u_m = 0 & \text{on } \partial\Omega \end{cases}$$

and

$$G(u) = \int_\Omega G(x, u).$$

If

$$|G(x,u)| \leqslant C \sum_{i=1}^{m} |u_i|^{s_i} \quad \forall (x,u) \in \Omega \times \mathbb{R}^m$$

for some $s_i \in (p_i, p_i^*)$, then

$$|G(u_\alpha)| \leqslant C \sum_{i=1}^{m} \alpha^{s_i/p_i} \|u_i\|_i^{s_i}$$

and hence (10.20) holds. Similarly, (10.27) holds if

$$|G(x,u)| \leqslant C \left(\sum_{i=1}^{m} |u_i|^{s_i} + 1 \right) \quad \forall (x,u) \in \Omega \times \mathbb{R}^m$$

for some $s_i \in (0, p_i)$.

10.3. Classification of Systems

The last section contains the critical group computations necessary to extend the existence and multiplicity results of Chapters 5 and 7 to systems. We classify (10.1) according to the growth of F as

(*i*) p-sublinear if

$$\lim_{\alpha \nearrow \infty} \frac{F(u_\alpha)}{\alpha} = 0, \quad \text{uniformly in } u \in \mathcal{M}, \tag{10.30}$$

(*ii*) asymptotically p-linear if

$$\lim_{\alpha \nearrow \infty} \frac{F(u_\alpha)}{\alpha} = \lambda\, J(u), \quad \text{uniformly in } u \in \mathcal{M} \tag{10.31}$$

for some $J \in \mathcal{J}$ with J' compact and $\lambda \neq 0$,

(*iii*) p-superlinear if

$$\lim_{\alpha \nearrow \infty} \frac{F(u_\alpha)}{\alpha} = \infty, \quad u \in \mathcal{M}. \tag{10.32}$$

Since (10.30) is the special case $\lambda = 0$ of (10.31), we treat the p-sublinear and asymptotically p-linear cases together. Critical groups of Φ at infinity are given by Proposition 10.14. In order to ensure that Φ satisfies the (PS) condition, we strengthen (10.31) by assuming (10.19) with

$$\lim_{\alpha \nearrow \infty} G'(u_\alpha)^\alpha = 0, \quad \text{uniformly in } u \in \mathcal{M} \tag{10.33}$$

in the notation of Section 10.1. Since

$$G(u_\alpha) - G(u_\beta) = \int_\beta^\alpha \frac{d}{dt}\,(G(u_t))\, dt = \int_\beta^\alpha \langle G'(u_t)^t, u/p \rangle\, dt$$

for $\alpha > \beta > 0$ by (10.8), then (10.27) holds and hence (10.31) follows from (10.11).

Example 10.16. In Example 10.15, for $\alpha > 0$,

$$\left\langle G'(u_\alpha)^\alpha, v\right\rangle = \sum_{i=1}^m \left\langle \alpha^{1/p_i - 1}\, G_{u_i}(u_\alpha), v_i\right\rangle_i = \sum_{i=1}^m \frac{1}{\alpha^{1-1/p_i}} \int_\Omega \frac{\partial G}{\partial u_i}(x, u_\alpha)\, v_i,$$

so if

$$\left|\frac{\partial G}{\partial u_i}\right| \leqslant C \left(\sum_{j=1}^m |u_j|^{s_{ij}-1} + 1\right) \qquad \forall (x,u) \in \Omega \times \mathbb{R}^m$$

for some $s_{ij} \in (1, 1 + p_j(p_i - 1)/p_i)$, then

$$\left\|G'(u_\alpha)^\alpha\right\|^* \leqslant C \sum_{i=1}^m \left(\sum_{j=1}^m \frac{\|u_j\|_j^{s_{ij}-1}}{\alpha^{1-1/p_i-(s_{ij}-1)/p_j}} + \frac{1}{\alpha^{1-1/p_i}}\right)$$

and hence (10.33) holds.

Lemma 10.17. *If* (10.19) *and* (10.33) *hold with* $\lambda \notin \sigma(A_p, J')$, *then* Φ *satisfies* (PS).

Proof. It is easy to see that Lemma 3.3 still holds, so it suffices to show that every (PS) sequence (u^j) is bounded, which is equivalent to the boundedness of $\alpha_j := I_p(u^j)$ by (10.3). If $\alpha_j \nearrow \infty$, set $\widetilde{u}^j := u^j_{\alpha_j^{-1}}$ and note that $I_p(\widetilde{u}^j) = 1$ by (10.10) and hence $\widetilde{u}^j \in \mathcal{M}$. We will show that a subsequence of $(\widetilde{u}^j)$ converges to an eigenvector $\widetilde{u}$ associated with λ, contradicting the assumption that $\lambda \notin \sigma(A_p, J')$.

By (10.19),

$$A_p\, u^j - \lambda\, J'(u^j) = \Phi'(u^j) - G'(u^j),$$

and hence

$$(A_p\, u^j)^{\alpha_j} - \lambda\, J'(u^j)^{\alpha_j} = \Phi'(u^j)^{\alpha_j} - G'(u^j)^{\alpha_j}.$$

Since $u^j = \widetilde{u}^j_{\alpha_j}$,

$$(A_p\, u^j)^{\alpha_j} = A_p\, \widetilde{u}^j, \qquad J'(u^j)^{\alpha_j} = J'(\widetilde{u}^j), \qquad G'(u^j)^{\alpha_j} \to 0$$

by (10.9), Proposition 10.4 (*iii*), and (10.33), respectively. For j so large that $\alpha_j \geqslant 1$,

$$\left\|\Phi'(u^j)^{\alpha_j}\right\|^* = \left(\sum_{i=1}^m \left(\alpha_j^{1/p_i - 1} \left\|\Phi_{u_i}(u^j)\right\|_i^*\right)^2\right)^{\frac{1}{2}} \leqslant \frac{\left\|\Phi'(u^j)\right\|^*}{\alpha_j^{1-1/\min_i p_i}} \to 0.$$

Thus,

$$A_p\, \widetilde{u}^j - \lambda\, J'(\widetilde{u}^j) \to 0. \tag{10.34}$$

By (10.14), $(\widetilde{u}^j)$ is bounded, so a renamed subsequence converges weakly to some $\widetilde{u}$ since W is reflexive, and $J'(\widetilde{u}^j)$ converges in W^* for a further subsequence since J' is compact. Applying (10.34) to $\widetilde{u}^j - \widetilde{u}$ gives $\left\langle A_p\, \widetilde{u}^j, \widetilde{u}^j - \widetilde{u}\right\rangle \to 0$ since $\left\langle J'(\widetilde{u}^j), \widetilde{u}^j - \widetilde{u}\right\rangle \to 0$ by Lemma 3.4, so $\widetilde{u}^j \to \widetilde{u} \in \mathcal{M}$ for a subsequence by (A_4). Now passing to the limit in (10.34) gives $A_p\, \widetilde{u} = \lambda\, J'(\widetilde{u})$. □

The following existence result is now immediate from Propositions 10.14 and 3.16.

Theorem 10.18. *Assume* (A_{i1}) - (A_{i3}), (A_4), $J \in \mathcal{J}$ *with* J' *compact,* $\lambda \notin \sigma(A_p, J')$, *and* (10.19) *and* (10.33) *hold.*

(i) *If* $\lambda_1^- < \lambda < \lambda_1^+$, *then* Φ *has a global minimizer* u *with* $C^q(\Phi, u) \approx \delta_{q0}\, \mathbb{Z}_2$.

(ii) *If* $\lambda_{k+1}^- < \lambda < \lambda_k^-$ *or* $\lambda_k^+ < \lambda < \lambda_{k+1}^+$, *then* (10.1) *has a solution* u *with* $C^k(\Phi, u) \neq 0$.

Now suppose

$$F = \lambda_0\, J_0 - G_0 \tag{10.35}$$

where $J_0 \in \mathcal{J}$ with J_0' compact and

$$G_0(u_\alpha) = \mathrm{o}(\alpha) \text{ as } \alpha \searrow 0, \text{ uniformly in } u \in \mathcal{M}, \tag{10.36}$$

and let $\left(\lambda_{0k}^\pm\right)$ be the sequences of positive and negative eigenvalues associated with J_0.

Theorem 10.19. *Assume* (A_{i1}) - (A_{i3}), (A_4), $J_0, J \in \mathcal{J}$ *with* J_0', J' *compact,* $\lambda \notin \sigma(A_p, J')$, *and* (10.35), (10.36), (10.19), *and* (10.33) *hold.*

(i) *If* $\lambda_{0,k+1}^- < \lambda_0 < \lambda_{0k}^-$ *or* $\lambda_{0k}^+ < \lambda_0 < \lambda_{0,k+1}^+$, *and* $\lambda_1^- < \lambda < \lambda_1^+$, *then* (10.1) *has a solution* $u_1 \neq 0$. *If* $k \geqslant 2$, *then there is a second solution* $u_2 \neq 0$.

(ii) *If* $\lambda_{01}^- < \lambda_0 < \lambda_{01}^+$, *and* $\lambda_{k+1}^- < \lambda < \lambda_k^-$ *or* $\lambda_k^+ < \lambda < \lambda_{k+1}^+$, *then* (10.1) *has a mountain pass solution* $u \neq 0$ *with* $\Phi(u) > 0$.

PROOF. (*i*) Follows from Corollary 3.32 since $C^k(\Phi, 0) \neq 0$ by Proposition 10.13 and Φ is bounded from below by Proposition 10.14.

(*ii*) Follows from Corollary 3.31 since $C^0(\Phi, 0) \approx \mathbb{Z}_2$ by Proposition 10.13 and $C^0(\Phi, \infty) = 0$ by Proposition 10.14. □

In the p-superlinear case we also assume that $F(u)$ and $\langle F'(u), u/p - u/\mu\rangle$ are bounded from below and

$$H_\mu(u) := F(u) - \big\langle F'(u), u/\mu \big\rangle \tag{10.37}$$

is bounded from above for some $\mu = (\mu_1, \dots, \mu_m)$ with $\mu_i > p_i$ for each i. For $u \in \mathcal{M}$ and $\alpha > 0$,

$$\Phi(u_\alpha) = \alpha \left(1 - \frac{F(u_\alpha)}{\alpha}\right) \to -\infty \text{ as } \alpha \to \infty \tag{10.38}$$

by (10.10) and (10.32). By (10.38), (10.8), and (10.37),

$$\begin{aligned} \frac{d}{d\alpha}\big(\Phi(u_\alpha)\big) &= 1 - \alpha^{-1} \big\langle F'(u_\alpha), u_\alpha/p \big\rangle \\ &= \alpha^{-1} \big[\Phi(u_\alpha) - \big\langle F'(u_\alpha), u_\alpha/p - u_\alpha/\mu \big\rangle + H_\mu(u_\alpha)\big] \end{aligned} \tag{10.39}$$

and hence all critical values of Φ are greater than or equal to

$$a_0 := \inf \left[\langle F'(u), u/p - u/\mu \rangle - H_\mu(u) \right] \leqslant 0.$$

Lemma 10.20. *For each $a < a_0$, there is a C^1-map $A_a : \mathcal{M} \to (0, \infty)$ such that*

$$\Phi^a = \left\{ u_\alpha : u \in \mathcal{M},\, \alpha \geqslant A_a(u) \right\} \simeq \mathcal{M}.$$

PROOF. We have $\Phi(u_\alpha) \leqslant a$ for all sufficiently large $\alpha > 0$ by (10.38), and

$$\Phi(u_\alpha) \leqslant a \implies \frac{d}{d\alpha}\left(\Phi(u_\alpha)\right) < 0$$

by (10.39), so there is a unique $A_a(u) > 0$ such that

$$\alpha < \text{ (resp. } =, >) \; A_a(u) \implies \Phi(u_\alpha) > \text{ (resp. } =, <) \; a,$$

and the map A_a is C^1 by the implicit function theorem. Then $W \backslash \{0\}$, which is $\simeq \mathcal{M}$, deformation retracts to $\Phi^a = \left\{ u_\alpha : u \in \mathcal{M},\, \alpha \geqslant A_a(u) \right\}$ via

$$(W \backslash \{0\}) \times [0, 1] \to W \backslash \{0\},$$

$$(u, t) \mapsto \begin{cases} u_{1 - t + t\, \alpha_u^{-1} A_a(\widetilde{u})}, & u \in (W \backslash \{0\}) \backslash \Phi^a \\ u, & u \in \Phi^a \end{cases}$$

where $\alpha_u = I_p(u)$ and $\widetilde{u} = u_{\alpha_u^{-1}}$. □

Lemma 10.21. *Φ satisfies* (PS).

PROOF. If (u^j) is a (PS) sequence, then (A_{i2}) gives

$$\begin{aligned} c_0 \sum_{i=1}^m \left(\frac{1}{p_i} - \frac{1}{\mu_i} \right) \left\| u_i^j \right\|_i^{p_i} &\leqslant \sum_{i=1}^m \left(\frac{1}{p_i} - \frac{1}{\mu_i} \right) \left\langle A_{p_i} u_i^j, u_i^j \right\rangle_i \\ &= \Phi(u^j) - \left\langle \Phi'(u^j), u^j/\mu \right\rangle + H_\mu(u^j) \\ &\leqslant \mathrm{o}(\left\| u^j \right\|) + \mathrm{O}(1), \end{aligned}$$

so (u^j) is bounded and hence has a convergent subsequence. □

We can now prove

Theorem 10.22. *Assume* $\dim W = \infty$, (A_{i1}) - (A_{i3}) *and* (A_4) *hold,* $J_0 \in \mathcal{J}$ *with* J_0' *compact,* $\lambda_0 \notin \sigma(A_p, J_0')$, (10.35), (10.36), *and* (10.32) *hold, and* $F(u)$ *and* $\langle F'(u), u/p - u/\mu \rangle$ *are bounded from below and* $H_\mu(u)$ *is bounded from above for some* μ *with* $\mu_i > p_i$ *for each* i*. Then* (10.1) *has a solution* $u \neq 0$.

PROOF. By Proposition 10.13, $C^q(\Phi, 0) \neq 0$ for some q. Since any $a < a_0$ is less than all critical values and Φ is unbounded from below by (10.38),

$$\begin{aligned} C^q(\Phi, \infty) &\approx \widetilde{H}^{q-1}(\Phi^a) && \text{by Proposition 3.15 } (ii) \\ &\approx \widetilde{H}^{q-1}(\mathcal{M}) && \text{by Lemma 10.20} \\ &\approx \widetilde{H}^{q-1}(S) && \text{by Lemma 10.6} \\ &= 0 && \text{by Example 2.5.} \end{aligned}$$

So the conclusion follows from Proposition 3.28 (*ii*). □

Example 10.23. In problem (10.5), let $\mu = (\mu_1, \ldots, \mu_m)$ with $\mu_i > p_i$ for each i, set

$$r_\mu(u) := \sum_{i=1}^m |u_i|^{\mu_i},$$

and assume that $\nabla_u F(x, u) \cdot (u/p - u/\mu)$ is bounded from below and

$$(10.40) \qquad r_\mu(u) \geqslant R \implies 0 < F(x, u) \leqslant \nabla_u F(x, u) \cdot u/\mu$$

for some $R > 0$. Since $\nabla_u F(x, u)$, and hence also $F(x, u)$, is bounded on bounded sets by (10.6), it then follows that $F(u)$ and

$$\langle F'(u), u/p - u/\mu \rangle = \int_\Omega \nabla_u F(x, u) \cdot (u/p - u/\mu)$$

are bounded from below. The group $G = \mathbb{R} \setminus \{0\}$ also has the representation

$$\rho_{1/\mu}(\alpha)\, u = (|\alpha|^{1/\mu_1 - 1}\, \alpha\, u_1, \ldots, |\alpha|^{1/\mu_m - 1}\, \alpha\, u_m)$$

on R^m. If $r_\mu(u) \geqslant R$, set $\alpha_u := r_\mu(u)/R \geqslant 1$ and $\widetilde{u} := \rho_{1/\mu}(\alpha_u^{-1})\, u$, and note that $r_\mu(\widetilde{u}) = \alpha_u^{-1}\, r_\mu(u) = R$. Then for $\alpha \geqslant 1$, $r_\mu(\rho_{1/\mu}(\alpha)\, \widetilde{u}) = \alpha\, r_\mu(\widetilde{u}) = \alpha\, R \geqslant R$ and hence

$$\begin{aligned} \frac{d}{d\alpha} \big(F(x, \rho_{1/\mu}(\alpha)\, \widetilde{u}) \big) &= \sum_{i=1}^m \frac{\partial F}{\partial u_i}(x, \rho_{1/\mu}(\alpha)\, \widetilde{u})\, \alpha^{1/\mu_i - 1}\, \widetilde{u}_i/\mu_i \\ &= \alpha^{-1}\, \nabla_u F(x, \rho_{1/\mu}(\alpha)\, \widetilde{u}) \cdot \rho_{1/\mu}(\alpha)\, \widetilde{u}/\mu \geqslant \alpha^{-1}\, F(x, \rho_{1/\mu}(\alpha)\, \widetilde{u}) \end{aligned}$$

by (10.40), and integrating this from $\alpha = 1$ to α_u gives

$$F(x, u) \geqslant F(x, \widetilde{u})\, \alpha_u.$$

So

$$F(x, u) \geqslant c(x)\, r_\mu(u) - C \quad \forall (x, u)$$

where

$$c(x) = \min_{r_\mu(u) = R} \frac{F(x, u)}{R} > 0$$

and $C > 0$, and hence

$$\frac{F(u_\alpha)}{\alpha} \geqslant \sum_{i=1}^m \alpha^{\mu_i/p_i - 1} \int_\Omega c(x)\, |u_i|^{\mu_i} - \frac{C\, |\Omega|}{\alpha} \to \infty, \qquad u \in \mathcal{M}.$$

Finally, since

$$H_\mu(x,u) := F(x,u) - \nabla_u F(x,u) \cdot u/\mu$$

is $\leqslant 0$ for $r_\mu(u) \geqslant R$ by (10.40) and bounded for $r_\mu(u) < R$,

$$H_\mu(u) = \int_\Omega H_\mu(x,u)$$

is bounded from above.

Bibliography

[1] Ravi Agarwal, Martin Bohner, Donal O'Regan, and Allan Peterson. Dynamic equations on time scales: a survey. *J. Comput. Appl. Math.*, 141(1-2):1–26, 2002. Dynamic equations on time scales.

[2] Ravi P. Agarwal, Victoria Otero-Espinar, Kanishka Perera, and Dolores R. Vivero. Basic properties of Sobolev's spaces on time scales. *Adv. Difference Equ.*, pages Art. ID 38121, 14, 2006.

[3] S. Ahmad, A. C. Lazer, and J. L. Paul. Elementary critical point theory and perturbations of elliptic boundary value problems at resonance. *Indiana Univ. Math. J.*, 25(10):933–944, 1976.

[4] H. Amann and E. Zehnder. Nontrivial solutions for a class of nonresonance problems and applications to nonlinear differential equations. *Ann. Scuola Norm. Sup. Pisa Cl. Sci. (4)*, 7(4):539–603, 1980.

[5] Antonio Ambrosetti. Differential equations with multiple solutions and nonlinear functional analysis. In *Equadiff 82 (Würzburg, 1982)*, volume 1017 of *Lecture Notes in Math.*, pages 10–37. Springer, Berlin, 1983.

[6] Antonio Ambrosetti. Elliptic equations with jumping nonlinearities. *J. Math. Phys. Sci.*, 18(1):1–12, 1984.

[7] Antonio Ambrosetti and Paul H. Rabinowitz. Dual variational methods in critical point theory and applications. *J. Functional Analysis*, 14:349–381, 1973.

[8] A. Anane and N. Tsouli. On the second eigenvalue of the p-Laplacian. In *Nonlinear partial differential equations (Fès, 1994)*, volume 343 of *Pitman Res. Notes Math. Ser.*, pages 1–9. Longman, Harlow, 1996.

[9] Aomar Anane. Simplicité et isolation de la première valeur propre du p-laplacien avec poids. *C. R. Acad. Sci. Paris Sér. I Math.*, 305(16):725–728, 1987.

[10] A. Ayoujil and A. R. El Amrouss. Existence of multiple solutions to elliptic equations. *Int. J. Math. Stat.*, 2(S08):56–61, 2008.

[11] A. Ayoujil and A. R. El Amrouss. Multiplicity results for quasi-linear problems. *Nonlinear Anal.*, 68(7):1802–1815, 2008.

[12] Abdesslem Ayoujil and Abdel R. El Amrouss. Nontrivial solutions for nonlinear elliptic problems via Morse theory. In *Proceedings of the 2005 Oujda International Conference on Nonlinear Analysis*, volume 14 of *Electron. J. Differ. Equ. Conf.*, pages 125–133 (electronic), San Marcos, TX, 2006. Southwest Texas State Univ.

[13] P. Bartolo, V. Benci, and D. Fortunato. Abstract critical point theorems and applications to some nonlinear problems with "strong" resonance at infinity. *Nonlinear Anal.*, 7(9):981–1012, 1983.

[14] Thomas Bartsch and Shujie Li. Critical point theory for asymptotically quadratic functionals and applications to problems with resonance. *Nonlinear Anal.*, 28(3):419–441, 1997.

[15] Thomas Bartsch and Zhaoli Liu. Location and critical groups of critical points in Banach spaces with an application to nonlinear eigenvalue problems. *Adv. Differential Equations*, 9(5-6):645–676, 2004.

[16] Vieri Benci. On critical point theory for indefinite functionals in the presence of symmetries. *Trans. Amer. Math. Soc.*, 274(2):533–572, 1982.

[17] Vieri Benci. Some applications of the generalized Morse-Conley index. *Confer. Sem. Mat. Univ. Bari*, (218):32, 1987.

[18] Vieri Benci. A new approach to the Morse-Conley theory and some applications. *Ann. Mat. Pura Appl. (4)*, 158:231–305, 1991.

[19] Vieri Benci. Introduction to Morse theory: a new approach. In *Topological nonlinear analysis*, volume 15 of *Progr. Nonlinear Differential Equations Appl.*, pages 37–177. Birkhäuser Boston, Boston, MA, 1995.

[20] Vieri Benci and Paul H. Rabinowitz. Critical point theorems for indefinite functionals. *Invent. Math.*, 52(3):241–273, 1979.

[21] Haïm Brezis and Louis Nirenberg. Remarks on finding critical points. *Comm. Pure Appl. Math.*, 44(8-9):939–963, 1991.

[22] Nguyên Phuong Các. On nontrivial solutions of a Dirichlet problem whose jumping nonlinearity crosses a multiple eigenvalue. *J. Differential Equations*, 80(2):379–404, 1989.

[23] Alberto Cambini. Sul lemma di M. Morse. *Boll. Un. Mat. Ital. (4)*, 7:87–93, 1973.

[24] A. Castro and A. C. Lazer. Applications of a maximin principle. *Rev. Colombiana Mat.*, 10:141–149, 1976.

[25] Giovanna Cerami. An existence criterion for the critical points on unbounded manifolds. *Istit. Lombardo Accad. Sci. Lett. Rend. A*, 112(2):332–336 (1979), 1978.

[26] K. C. Chang and N. Ghoussoub. The Conley index and the critical groups via an extension of Gromoll-Meyer theory. *Topol. Methods Nonlinear Anal.*, 7(1):77–93, 1996.

[27] K. C. Chang. Solutions of asymptotically linear operator equations via Morse theory. *Comm. Pure Appl. Math.*, 34(5):693–712, 1981.

[28] K. C. Chang. *Infinite-dimensional Morse theory and multiple solution problems*, volume 6 of *Progress in Nonlinear Differential Equations and their Applications*. Birkhäuser Boston Inc., Boston, MA, 1993.

[29] S. Cingolani and G. Vannella. Morse index computations for a class of functionals defined in Banach spaces. In *Nonlinear equations: methods, models and applications (Bergamo, 2001)*, volume 54 of *Progr. Nonlinear Differential Equations Appl.*, pages 107–116. Birkhäuser, Basel, 2003.

[30] Silvia Cingolani and Marco Degiovanni. Nontrivial solutions for p-Laplace equations with right-hand side having p-linear growth at infinity. *Comm. Partial Differential Equations*, 30(7-9):1191–1203, 2005.

[31] Silvia Cingolani and Giuseppina Vannella. Some results on critical groups for a class of functionals defined on Sobolev Banach spaces. *Atti Accad. Naz. Lincei Cl. Sci. Fis. Mat. Natur. Rend. Lincei (9) Mat. Appl.*, 12:199–203 (2002), 2001.

[32] J.-N. Corvellec. A general approach to the min-max principle. *Z. Anal. Anwendungen*, 16(2):405–433, 1997.

[33] J.-N. Corvellec and A. Hantoute. Homotopical stability of isolated critical points of continuous functionals. *Set-Valued Anal.*, 10(2-3):143–164, 2002. Calculus of variations, nonsmooth analysis and related topics.

[34] Jean-Noël Corvellec, Marco Degiovanni, and Marco Marzocchi. Deformation properties for continuous functionals and critical point theory. *Topol. Methods Nonlinear Anal.*, 1(1):151–171, 1993.

[35] M. Cuesta. On the Fučík spectrum of the Laplacian and the p-Laplacian. In *Proceedings of Seminar in Differential Equations*, pages 67–96, Kvilda, Czech Republic, May 29 – June 2, 2000. Centre of Applied Mathematics, Faculty of Applied Sciences, University of West Bohemia in Pilsen.

[36] M. Cuesta, D. de Figueiredo, and J.-P. Gossez. The beginning of the Fučik spectrum for the p-Laplacian. *J. Differential Equations*, 159(1):212–238, 1999.

[37] Mabel Cuesta and Jean-Pierre Gossez. A variational approach to nonresonance with respect to the Fučik spectrum. *Nonlinear Anal.*, 19(5):487–500, 1992.

[38] E. N. Dancer. On the Dirichlet problem for weakly non-linear elliptic partial differential equations. *Proc. Roy. Soc. Edinburgh Sect. A*, 76(4):283–300, 1976/77.

[39] E. N. Dancer. Corrigendum: "On the Dirichlet problem for weakly nonlinear elliptic partial differential equations" [Proc. Roy. Soc. Edinburgh Sect. A **76** (1976/77), no. 4, 283–300; MR **58** #17506]. *Proc. Roy. Soc. Edinburgh Sect. A*, 89(1-2):15, 1981.

[40] Norman Dancer and Kanishka Perera. Some remarks on the Fučík spectrum of the p-Laplacian and critical groups. *J. Math. Anal. Appl.*, 254(1):164–177, 2001.

[41] D. G. de Figueiredo and J.-P. Gossez. On the first curve of the Fučik spectrum of an elliptic operator. *Differential Integral Equations*, 7(5-6):1285–1302, 1994.

[42] M. Degiovanni, S. Lancelotti, and K. Perera. Nontrivial solutions of p-superlinear p-laplacian problems via a cohomological local splitting. Commun. Contemp. Math., to appear.

[43] Marco Degiovanni and Sergio Lancelotti. Linking over cones and nontrivial solutions for p-Laplace equations with p-superlinear nonlinearity. 2007.

[44] Marco Degiovanni and Sergio Lancelotti. Linking solutions for p-Laplace equations with nonlinearity at critical growth. *J. Funct. Anal.*, 256(11):3643–3659, 2009.

[45] Albrecht Dold. Partitions of unity in the theory of fibrations. *Ann. of Math. (2)*, 78:223–255, 1963.

[46] P. Drábek. *Solvability and bifurcations of nonlinear equations*, volume 264 of *Pitman Research Notes in Mathematics Series*. Longman Scientific & Technical, Harlow, 1992.

[47] Pavel Drábek and Stephen B. Robinson. Resonance problems for the p-Laplacian. *J. Funct. Anal.*, 169(1):189–200, 1999.

[48] Juan R. Esteban and Juan L. Vázquez. On the equation of turbulent filtration in one-dimensional porous media. *Nonlinear Anal.*, 10(11):1303–1325, 1986.

[49] Edward R. Fadell and Paul H. Rabinowitz. Generalized cohomological index theories for Lie group actions with an application to bifurcation questions for Hamiltonian systems. *Invent. Math.*, 45(2):139–174, 1978.

[50] Fei Fang and Shibo Liu. Nontrivial solutions of superlinear p-Laplacian equations. *J. Math. Anal. Appl.*, 351(1):138–146, 2009.

[51] Svatopluk Fučík. Boundary value problems with jumping nonlinearities. *Časopis Pěst. Mat.*, 101(1):69–87, 1976.

[52] N. Ghoussoub. Location, multiplicity and Morse indices of min-max critical points. *J. Reine Angew. Math.*, 417:27–76, 1991.

[53] Nassif Ghoussoub. *Duality and perturbation methods in critical point theory*, volume 107 of *Cambridge Tracts in Mathematics*. Cambridge University Press, Cambridge, 1993. With appendices by David Robinson.

[54] Detlef Gromoll and Wolfgang Meyer. On differentiable functions with isolated critical points. *Topology*, 8:361–369, 1969.

[55] Yuxia Guo and Jiaquan Liu. Solutions of p-sublinear p-Laplacian equation via Morse theory. *J. London Math. Soc. (2)*, 72(3):632–644, 2005.

[56] Helmut Hofer. The topological degree at a critical point of mountain-pass type. 45:501–509, 1986.

[57] Mei-Yue Jiang. Critical groups and multiple solutions of the p-Laplacian equations. *Nonlinear Anal.*, 59(8):1221–1241, 2004.

[58] Quansen Jiu and Jiabao Su. Existence and multiplicity results for Dirichlet problems with p-Laplacian. *J. Math. Anal. Appl.*, 281(2):587–601, 2003.

[59] John L. Kelley. *General topology*. Springer-Verlag, New York, 1975. Reprint of the 1955 edition [Van Nostrand, Toronto, Ont.], Graduate Texts in Mathematics, No. 27.

[60] M. A. Krasnosel'skii. *Topological methods in the theory of nonlinear integral equations*. Translated by A. H. Armstrong; translation edited by J. Burlak. A Pergamon Press Book. The Macmillan Co., New York, 1964.

[61] Wojciech Kryszewski and Andrzej Szulkin. An infinite-dimensional Morse theory with applications. *Trans. Amer. Math. Soc.*, 349(8):3181–3234, 1997.

[62] Nicolaas H. Kuiper. C^1-equivalence of functions near isolated critical points. pages 199–218. Ann. of Math. Studies, No. 69, 1972.

[63] Sergio Lancelotti. Existence of nontrivial solutions for semilinear problems with strictly differentiable nonlinearity. *Abstr. Appl. Anal.*, pages Art. ID 62458, 14, 2006.

[64] Serge Lang. *Fundamentals of differential geometry*, volume 191 of *Graduate Texts in Mathematics*. Springer-Verlag, New York, 1999.

[65] A. C. Lazer and S. Solimini. Nontrivial solutions of operator equations and Morse indices of critical points of min-max type. *Nonlinear Anal.*, 12(8):761–775, 1988.

[66] Shu Jie Li and J. Q. Liu. Morse theory and asymptotic linear Hamiltonian system. *J. Differential Equations*, 78(1):53–73, 1989.

[67] Shu Jie Li and Michel Willem. Applications of local linking to critical point theory. *J. Math. Anal. Appl.*, 189(1):6–32, 1995.

[68] Peter Lindqvist. On the equation $\operatorname{div}(|\nabla u|^{p-2}\nabla u) + \lambda|u|^{p-2}u = 0$. *Proc. Amer. Math. Soc.*, 109(1):157–164, 1990.

[69] Peter Lindqvist. Addendum: "On the equation $\operatorname{div}(|\nabla u|^{p-2}\nabla u) + \lambda|u|^{p-2}u = 0$" [Proc. Amer. Math. Soc. 109 (1990), no. 1, 157–164; MR 90h:35088]. *Proc. Amer. Math. Soc.*, 116(2):583–584, 1992.

[70] Disheng Liu and Di Geng. Infinitely many solutions for the p-Laplace equations with nonsymmetric perturbations. *Electron. J. Differential Equations*, pages No. 101, 7, 2008.

[71] Jia Quan Liu. The Morse index of a saddle point. *Systems Sci. Math. Sci.*, 2(1):32–39, 1989.

[72] Jia Quan Liu and Shu Jie Li. An existence theorem for multiple critical points and its application. *Kexue Tongbao (Chinese)*, 29(17):1025–1027, 1984.

[73] Jiaquan Liu and Shibo Liu. The existence of multiple solutions to quasilinear elliptic equations. *Bull. London Math. Soc.*, 37(4):592–600, 2005.

[74] Jiaquan Liu and Jiabao Su. Remarks on multiple nontrivial solutions for quasi-linear resonant problems. *J. Math. Anal. Appl.*, 258(1):209–222, 2001.

[75] Shibo Liu and Shujie Li. Existence of solutions for asymptotically 'linear' p-Laplacian equations. *Bull. London Math. Soc.*, 36(1):81–87, 2004.

[76] L. Ljusternik and L. Schnirelmann. *Methodes topologique dans les problémes variationnels*. Hermann and Cie, Paris, 1934.

[77] Caryl Ann Margulies and William Margulies. An example of the Fučik spectrum. *Nonlinear Anal.*, 29(12):1373–1378, 1997.

[78] A. Marino and G. Prodi. Metodi perturbativi nella teoria di Morse. *Boll. Un. Mat. Ital. (4)*, 11(3, suppl.):1–32, 1975. Collection of articles dedicated to Giovanni Sansone on the occasion of his eighty-fifth birthday.

[79] Antonio Marino and Giovanni Prodi. La teoria di Morse per gli spazi di Hilbert. *Rend. Sem. Mat. Univ. Padova*, 41:43–68, 1968.

[80] J. Mawhin and M. Willem. On the generalized Morse lemma. *Bull. Soc. Math. Belg. Sér. B*, 37(2):23–29, 1985.

[81] Jean Mawhin and Michel Willem. *Critical point theory and Hamiltonian systems*, volume 74 of *Applied Mathematical Sciences*. Springer-Verlag, New York, 1989.

[82] Everaldo Medeiros and Kanishka Perera. Multiplicity of solutions for a quasilinear elliptic problem via the cohomological index. *Nonlinear Anal.*, 71(9):3654–3660, 2009.

[83] Anna Maria Micheletti and Angela Pistoia. On the Fučík spectrum for the p-Laplacian. *Differential Integral Equations*, 14(7):867–882, 2001.

[84] J. Milnor. Morse theory. pages vi+153, 1963.

[85] Marston Morse. Relations between the critical points of a real function of n independent variables. *Trans. Amer. Math. Soc.*, 27(3):345–396, 1925.

[86] Dumitru Motreanu and Kanishka Perera. Multiple nontrivial solutions of Neumann p-Laplacian systems. *Topol. Methods Nonlinear Anal.*, 34(1):41–48, 2009.

[87] Wei Ming Ni. Some minimax principles and their applications in nonlinear elliptic equations. *J. Analyse Math.*, 37:248–275, 1980.

[88] L. Nirenberg. Topics in nonlinear functional analysis. pages viii+259, 1974. With a chapter by E. Zehnder, Notes by R. A. Artino, Lecture Notes, 1973–1974.

[89] L. Nirenberg. Variational and topological methods in nonlinear problems. *Bull. Amer. Math. Soc. (N.S.)*, 4(3):267–302, 1981.

[90] Juan Francisco Padial, Peter Takáč, and Lourdes Tello. An antimaximum principle for a degenerate parabolic problem. In *Ninth International Conference Zaragoza-Pau on Applied Mathematics and Statistics*, volume 33 of *Monogr. Semin. Mat. García Galdeano*, pages 433–440. Prensas Univ. Zaragoza, Zaragoza, 2006.

[91] R. S. Palais and S. Smale. A generalized Morse theory. *Bull. Amer. Math. Soc.*, 70:165–172, 1964.

[92] Richard S. Palais. Morse theory on Hilbert manifolds. *Topology*, 2:299–340, 1963.

[93] Richard S. Palais. Critical point theory and the minimax principle. In *Global Analysis (Proc. Sympos. Pure Math., Vol. XV, Berkeley, Calif, 1968)*, pages 185–212. Amer. Math. Soc., Providence, R.I., 1970.

[94] Kanishka Perera. Critical groups of pairs of critical points produced by linking subsets. *J. Differential Equations*, 140(1):142–160, 1997.

[95] Kanishka Perera. Critical groups of critical points produced by local linking with applications. *Abstr. Appl. Anal.*, 3(3-4):437–446, 1998.

[96] Kanishka Perera. Homological local linking. *Abstr. Appl. Anal.*, 3(1-2):181–189, 1998.

[97] Kanishka Perera. One-sided resonance for quasilinear problems with asymmetric nonlinearities. *Abstr. Appl. Anal.*, 7(1):53–60, 2002.

[98] Kanishka Perera. Nontrivial critical groups in p-Laplacian problems via the Yang index. *Topol. Methods Nonlinear Anal.*, 21(2):301–309, 2003.

[99] Kanishka Perera. Nontrivial solutions of p-superlinear p-Laplacian problems. *Appl. Anal.*, 82(9):883–888, 2003.

[100] Kanishka Perera. p-superlinear problems with jumping nonlinearities. In *Nonlinear analysis and applications: to V. Lakshmikantham on his 80th birthday. Vol. 1, 2*, pages 823–829. Kluwer Acad. Publ., Dordrecht, 2003.

[101] Kanishka Perera. On the Fučík spectrum of the p-Laplacian. *NoDEA Nonlinear Differential Equations Appl.*, 11(2):259–270, 2004.

[102] Kanishka Perera and Martin Schechter. Morse index estimates in saddle point theorems without a finite-dimensional closed loop. *Indiana Univ. Math. J.*, 47(3):1083–1095, 1998.

[103] Kanishka Perera and Martin Schechter. Critical groups in saddle point theorems without a finite dimensional closed loop. *Math. Nachr.*, 243:156–164, 2002.

[104] Kanishka Perera and Martin Schechter. Double resonance problems with respect to the Fučík spectrum. *Indiana Univ. Math. J.*, 52(1):1–17, 2003.

[105] Kanishka Perera and Andrzej Szulkin. p-Laplacian problems where the nonlinearity crosses an eigenvalue. *Discrete Contin. Dyn. Syst.*, 13(3):743–753, 2005.

[106] Everett Pitcher. Inequalities of critical point theory. *Bull. Amer. Math. Soc.*, 64(1):1–30, 1958.

[107] Gui Jie Qi. Extension of Mountain Pass Lemma. *Kexue Tongbao (English Ed.)*, 32(12):798–801, 1987.

[108] P. H. Rabinowitz. Variational methods for nonlinear eigenvalue problems. pages 139–195, 1974.

[109] Paul H. Rabinowitz. Some critical point theorems and applications to semilinear elliptic partial differential equations. *Ann. Scuola Norm. Sup. Pisa Cl. Sci. (4)*, 5(1):215–223, 1978.

[110] Paul H. Rabinowitz. Some minimax theorems and applications to nonlinear partial differential equations. In *Nonlinear analysis (collection of papers in honor of Erich H. Rothe)*, pages 161–177. Academic Press, New York, 1978.

[111] Paul H. Rabinowitz. *Minimax methods in critical point theory with applications to differential equations*, volume 65 of *CBMS Regional Conference Series in Mathematics*. Published for the Conference Board of the Mathematical Sciences, Washington, DC, 1986.

[112] Miguel Ramos and Luis Sanchez. Homotopical linking and Morse index estimates in min-max theorems. *Manuscripta Math.*, 87(3):269–284, 1995.

[113] N. K. Ribarska, Ts. Y. Tsachev, and M. I. Krastanov. A saddle point theorem without a finite-dimensional closed loop. *C. R. Acad. Bulgare Sci.*, 51(11-12):13–16, 1998.

[114] E. H. Rothe. Some remarks on critical point theory in Hilbert space. In *Nonlinear Problems (Proc. Sympos., Madison, Wis., (1962)*, pages 233–256. Univ. of Wisconsin Press, Madison, Wis., 1963.

[115] E. H. Rothe. Some remarks on critical point theory in Hilbert space (continuation). *J. Math. Anal. Appl.*, 20:515–520, 1967.

[116] E. H. Rothe. On continuity and approximation questions concerning critical Morse groups in Hilbert space. In *Symposium on Infinite-Dimensional Topology (Louisiana State Univ., Baton Rouge, La., 1967)*, pages 275–295. Ann. of Math. Studies, No. 69. Princeton Univ. Press, Princeton, N. J., 1972.

[117] E. H. Rothe. Morse theory in Hilbert space. *Rocky Mountain J. Math.*, 3:251–274, 1973. Rocky Mountain Consortium Symposium on Nonlinear Eigenvalue Problems (Santa Fe, N.M., 1971).

[118] Martin Schechter. Splitting subspaces and saddle points. *Appl. Anal.*, 49(1-2):33–48, 1993.

[119] Martin Schechter. The Fučík spectrum. *Indiana Univ. Math. J.*, 43(4):1139–1157, 1994.

[120] Martin Schechter. New linking theorems. *Rend. Sem. Mat. Univ. Padova*, 99:255–269, 1998.

[121] Martin Schechter. *Linking methods in critical point theory*. Birkhäuser Boston Inc., Boston, MA, 1999.

[122] Martin Schechter and Kyril Tintarev. Pairs of critical points produced by linking subsets with applications to semilinear elliptic problems. *Bull. Soc. Math. Belg. Sér. B*, 44(3):249–261, 1992.

[123] J. T. Schwartz. Nonlinear functional analysis. pages vii+236, 1969. Notes by H. Fattorini, R. Nirenberg and H. Porta, with an additional chapter by Hermann Karcher, Notes on Mathematics and its Applications.

[124] S. Smale. Morse theory and a non-linear generalization of the Dirichlet problem. *Ann. of Math. (2)*, 80:382–396, 1964.

[125] Sergio Solimini. Morse index estimates in min-max theorems. *Manuscripta Math.*, 63(4):421–453, 1989.

[126] Edwin H. Spanier. *Algebraic topology*. Springer-Verlag, New York, 1994. Corrected reprint of the 1966 original.

[127] Norman Steenrod. *The topology of fibre bundles*. Princeton Landmarks in Mathematics. Princeton University Press, Princeton, NJ, 1999. Reprint of the 1957 edition, Princeton Paperbacks.

[128] Andrzej Szulkin. Cohomology and Morse theory for strongly indefinite functionals. *Math. Z.*, 209(3):375–418, 1992.

[129] Mieko Tanaka. On the existence of a non-trivial solution for the p-Laplacian equation with a jumping nonlinearity. *Tokyo J. Math.*, 31(2):333–341, 2008.

[130] Giuseppina Vannella. Critical groups estimates for a soliton type problem. In *Proceedings of the Third World Congress of Nonlinear Analysts, Part 9 (Catania, 2000)*, volume 47, pages 5999–6008, 2001.

[131] Zhi Qiang Wang. A note on the second variation theorem. *Acta Math. Sinica*, 30(1):106–110, 1987.

[132] Zhi Qiang Wang. On a superlinear elliptic equation. *Ann. Inst. H. Poincaré Anal. Non Linéaire*, 8(1):43–57, 1991.

[133] Chung-Tao Yang. On theorems of Borsuk-Ulam, Kakutani-Yamabe-Yujobô and Dyson. I. *Ann. of Math. (2)*, 60:262–282, 1954.

Titles in This Series

161 **Kanishka Perera, Ravi P. Agarwal, and Donal O'Regan,** Morse theoretic aspects of p-Laplacian type operators, 2010

160 **Alexander S. Kechris,** Global aspects of ergodic group actions, 2010

159 **Matthew Baker and Robert Rumely,** Potential theory and dynamics on the Berkovich projective line, 2010

158 **D. R. Yafaev,** Mathematical scattering theory: Analytic theory, 2010

157 **Xia Chen,** Random walk intersections: Large deviations and related topics, 2010

156 **Jaime Angulo Pava,** Nonlinear dispersive equations: Existence and stability of solitary and periodic travelling wave solutions, 2009

155 **Yiannis N. Moschovakis,** Descriptive set theory, 2009

154 **Andreas Čap and Jan Slovák,** Parabolic geometries I: Background and general theory, 2009

153 **Habib Ammari, Hyeonbae Kang, and Hyundae Lee,** Layer potential techniques in spectral analysis, 2009

152 **János Pach and Micha Sharir,** Combinatorial geometry and its algorithmic applications: The Alcála lectures, 2009

151 **Ernst Binz and Sonja Pods,** The geometry of Heisenberg groups: With applications in signal theory, optics, quantization, and field quantization, 2008

150 **Bangming Deng, Jie Du, Brian Parshall, and Jianpan Wang,** Finite dimensional algebras and quantum groups, 2008

149 **Gerald B. Folland,** Quantum field theory: A tourist guide for mathematicians, 2008

148 **Patrick Dehornoy with Ivan Dynnikov, Dale Rolfsen, and Bert Wiest,** Ordering braids, 2008

147 **David J. Benson and Stephen D. Smith,** Classifying spaces of sporadic groups, 2008

146 **Murray Marshall,** Positive polynomials and sums of squares, 2008

145 **Tuna Altinel, Alexandre V. Borovik, and Gregory Cherlin,** Simple groups of finite Morley rank, 2008

144 **Bennett Chow, Sun-Chin Chu, David Glickenstein, Christine Guenther, James Isenberg, Tom Ivey, Dan Knopf, Peng Lu, Feng Luo, and Lei Ni,** The Ricci flow: Techniques and applications, Part II: Analytic aspects, 2008

143 **Alexander Molev,** Yangians and classical Lie algebras, 2007

142 **Joseph A. Wolf,** Harmonic analysis on commutative spaces, 2007

141 **Vladimir Maz′ya and Gunther Schmidt,** Approximate approximations, 2007

140 **Elisabetta Barletta, Sorin Dragomir, and Krishan L. Duggal,** Foliations in Cauchy-Riemann geometry, 2007

139 **Michael Tsfasman, Serge Vlăduţ, and Dmitry Nogin,** Algebraic geometric codes: Basic notions, 2007

138 **Kehe Zhu,** Operator theory in function spaces, 2007

137 **Mikhail G. Katz,** Systolic geometry and topology, 2007

136 **Jean-Michel Coron,** Control and nonlinearity, 2007

135 **Bennett Chow, Sun-Chin Chu, David Glickenstein, Christine Guenther, James Isenberg, Tom Ivey, Dan Knopf, Peng Lu, Feng Luo, and Lei Ni,** The Ricci flow: Techniques and applications, Part I: Geometric aspects, 2007

134 **Dana P. Williams,** Crossed products of C^*-algebras, 2007

133 **Andrew Knightly and Charles Li,** Traces of Hecke operators, 2006

132 **J. P. May and J. Sigurdsson,** Parametrized homotopy theory, 2006

131 **Jin Feng and Thomas G. Kurtz,** Large deviations for stochastic processes, 2006

130 **Qing Han and Jia-Xing Hong,** Isometric embedding of Riemannian manifolds in Euclidean spaces, 2006

129 **William M. Singer,** Steenrod squares in spectral sequences, 2006

128 **Athanassios S. Fokas, Alexander R. Its, Andrei A. Kapaev, and Victor Yu. Novokshenov,** Painlevé transcendents, 2006

127 **Nikolai Chernov and Roberto Markarian,** Chaotic billiards, 2006

126 **Sen-Zhong Huang,** Gradient inequalities, 2006

125 **Joseph A. Cima, Alec L. Matheson, and William T. Ross,** The Cauchy Transform, 2006

124 **Ido Efrat, Editor,** Valuations, orderings, and Milnor K-Theory, 2006

123 **Barbara Fantechi, Lothar Göttsche, Luc Illusie, Steven L. Kleiman, Nitin Nitsure, and Angelo Vistoli,** Fundamental algebraic geometry: Grothendieck's FGA explained, 2005

122 **Antonio Giambruno and Mikhail Zaicev, Editors,** Polynomial identities and asymptotic methods, 2005

121 **Anton Zettl,** Sturm-Liouville theory, 2005

120 **Barry Simon,** Trace ideals and their applications, 2005

119 **Tian Ma and Shouhong Wang,** Geometric theory of incompressible flows with applications to fluid dynamics, 2005

118 **Alexandru Buium,** Arithmetic differential equations, 2005

117 **Volodymyr Nekrashevych,** Self-similar groups, 2005

116 **Alexander Koldobsky,** Fourier analysis in convex geometry, 2005

115 **Carlos Julio Moreno,** Advanced analytic number theory: L-functions, 2005

114 **Gregory F. Lawler,** Conformally invariant processes in the plane, 2005

113 **William G. Dwyer, Philip S. Hirschhorn, Daniel M. Kan, and Jeffrey H. Smith,** Homotopy limit functors on model categories and homotopical categories, 2004

112 **Michael Aschbacher and Stephen D. Smith,** The classification of quasithin groups II. Main theorems: The classification of simple QTKE-groups, 2004

111 **Michael Aschbacher and Stephen D. Smith,** The classification of quasithin groups I. Structure of strongly quasithin K-groups, 2004

110 **Bennett Chow and Dan Knopf,** The Ricci flow: An introduction, 2004

109 **Goro Shimura,** Arithmetic and analytic theories of quadratic forms and Clifford groups, 2004

108 **Michael Farber,** Topology of closed one-forms, 2004

107 **Jens Carsten Jantzen,** Representations of algebraic groups, 2003

106 **Hiroyuki Yoshida,** Absolute CM-periods, 2003

105 **Charalambos D. Aliprantis and Owen Burkinshaw,** Locally solid Riesz spaces with applications to economics, second edition, 2003

104 **Graham Everest, Alf van der Poorten, Igor Shparlinski, and Thomas Ward,** Recurrence sequences, 2003

103 **Octav Cornea, Gregory Lupton, John Oprea, and Daniel Tanré,** Lusternik-Schnirelmann category, 2003

102 **Linda Rass and John Radcliffe,** Spatial deterministic epidemics, 2003

101 **Eli Glasner,** Ergodic theory via joinings, 2003

100 **Peter Duren and Alexander Schuster,** Bergman spaces, 2004

99 **Philip S. Hirschhorn,** Model categories and their localizations, 2003

98 **Victor Guillemin, Viktor Ginzburg, and Yael Karshon,** Moment maps, cobordisms, and Hamiltonian group actions, 2002

97 **V. A. Vassiliev,** Applied Picard-Lefschetz theory, 2002

For a complete list of titles in this series, visit the
AMS Bookstore at **www.ams.org/bookstore/**.